T0206180

Smart Mini-Cameras

Smart
Mini-Cameras

Edited by
Tigran V. Galstian

CRC Press
Taylor & Francis Group
Boca Raton London New York

CRC Press is an imprint of the
Taylor & Francis Group, an **informa** business

CRC Press
Taylor & Francis Group
6000 Broken Sound Parkway NW, Suite 300
Boca Raton, FL 33487-2742

First issued in paperback 2020

© 2014 by Taylor & Francis Group, LLC
CRC Press is an imprint of Taylor & Francis Group, an Informa business

No claim to original U.S. Government works

Version Date: 20130722

ISBN 13: 978-0-367-57631-8 (pbk)
ISBN 13: 978-1-4665-1292-4 (hbk)

Library of Congress Cataloging-in-Publication Data

Smart mini-cameras / editor, Tigran V. Galstian.
       pages cm
    Includes bibliographical references and index.
    ISBN 978-1-4665-1292-4 (hardcover : alk. paper)
    1. Miniature cameras--Automatic control. 2. Miniature cameras--Design and construction. I. Galstian, Tigran, editor of compilation.

    TR260.5.S63 2014
    771.3'2--dc23   2013028853

Visit the Taylor & Francis Web site at
http://www.taylorandfrancis.com

and the CRC Press Web site at
http://www.crcpress.com

Contents

Foreword

Evolution of Miniature Cameras

In 2013, it is possible to buy a decent camera* for less than what a roll of film cost a generation ago. Mobile-phone cameras with 5 megapixels and autofocus are available for less than $5. Not only are these cameras very inexpensive, they are also very capable. Digital still cameras (DSC) *were* built to meet the demand for instant photography following up on an idea originally developed by Polaroid. Mobile-phone cameras took this a stage further, with images that can be "instantly everywhere."

How did this happen? Basic high-volume lens technology was developed through Kodak's Brownie in the 1930s to the Instamatic in the 1960s. The development of molded plastic lenses, again pioneered by Polaroid, made mass-produced optics cheaper and easier to make. Charge-coupled device (CCD) sensors, which could convert light into a matrix of electronic signals, began to replace film and its associated transport mechanisms. These were shortly followed by the more cost-effective complementary metal-oxide-semiconductor (CMOS) sensors.

In the early part of this century, the idea of putting a camera in a phone emerged. While this was made possible by the new, cheaper, CMOS image sensor and molded plastic lens technology, there was no certainty that this was something that people wanted. The initial idea seemed to be that people would take photographs and send them to each other via a multimedia messaging service (MMS). This was quite attractive to the carriers as, at the time, they charged for data by the bit or byte and the idea was seen as a revenue enhancer. The low quality of the images and the clunky interface required to send images limited the initial success of camera phones. However, the idea of always having a camera with you was attractive and began to take hold.

The advances in both the CMOS image sensor technology and the quality of the plastic lenses started to radically improve camera-phone image quality. This led to a virtuous cycle where, as the image quality improved, the cameras were used more, and as they were used more, the motivation to improve the image quality increased.

---

* Albeit a miniature camera module (MCM).

In tandem with this improved image quality and increased use, the means to share images became more compelling. Rather than send images by MMS, the mobile internet allowed people to post pictures on internet sites and social media. Sites such as YouTube gave a home on the internet for video which has now become central to the mobile imaging experience. This has had social consequences that would have been difficult for anyone to predict early on in the development of digital cameras. As cameras become increasingly capable and less expensive through the use of solid-state technologies, not just in the image sensor, but also in lensing and light capture, we can reasonably expect a trajectory similar to that seen in microprocessors and the emergence of an equivalent to Moore's law.

While it is difficult to predict the implications of this, it is always interesting to speculate. The written word became a unifying and important cultural phenomenon when books became easily reproducible. When the method of *creating* books also became reproducible with the advent of the typewriter and the word processor, the written word became an even stronger cultural and commercial factor. Some might say that this led directly to the internet as a vehicle to distribute the newfound plethora of written information. If we accept that man is primarily a visual creature, his very consciousness developing from the visual cortex, it is hardly surprising that a method highly convenient for making still and video images should be very successful. Once the internet was born, it was almost inevitable that we should want to populate it with images and video. The mobile-phone camera is a very appropriate tool for this.

Perhaps it is not too much of a stretch to speculate further. Could this cause as fundamental a change in the way we communicate and interact as was caused by the proliferation of writing? Images, especially video images, are more visceral than writing and as such may lead to more efficient communication. The use of video images in the "Arab Spring" certainly gave most of us a much more immediate and varied experience than we had been used to in the past. It was not just the medium, after all there had been plenty of video news reports in the past, but the fact that nearly every aspect of the situation could be shown because the mobile phone is almost undetectable and pretty much ubiquitous. Hopefully, this new, more immediate experience will allow all of us a greater degree of empathy and knowledge outside of our own personal experience. On the flipside, this is not to say that the puerile and the trivial will not make up the vast majority of material, as has been true of every medium in the past. It is also true that it will almost certainly lead to a greater loss of privacy.

Technologically speaking, predicting the future may be easier, less controversial, and more empirically based. As camera phones become increasingly capable, they are replacing and enhancing more and more devices and bring new utility. Already, there are "apps" that can scan and collate documents. These range from apps that are designed to keep track of business expenses to reading tax documents and all manner of archiving. Note-taking apps include facilities to photograph objects and documents, and to embed videos. Bar-code reading and object recognition are used to implement comparison shopping and are driving mobile commerce. Many of these uses of the camera phone were unimagined a few years ago. In the future, cameras may be used to track eye movements in order to control cursors, as range finders for automotive and aerospace applications, and as an integral part of artificial intelligence systems.

A particularly interesting emerging photographic aspect of mobile imaging, which may well be a harbinger of the future, is the invention of the array camera. While we can confidently expect known features such as zoom and optical image stabilization to be quickly incorporated into mobile cameras, array cameras may take photography to a different level. An interesting initial application of this is the ARGUS-IS drone camera (the military always get to develop the best toys!), which is an array of 368 standard mobile-phone cameras, reportedly. These generate a 1.8-gigapixel image, allowing an incredible amount of detail across a very large horizon to be collected. It is not difficult to imagine that in the future the very high-end cameras, the equivalent of digital single-lens reflex (SLR) cameras and even larger format cameras used for making movies, will be large arrays of cameras, using the miniature camera as a building block. In the mobile phone, one can envision a low-end phone having a "unicamera," made of one building block and a higher-end phone using a "multicamera," a small array of the same cameras. It is almost certain that the very fact that it is easy to record videos and images and to know the position and vantage points of such images will lead to ever more accurate models of reality.

However, the reader's speculation on these matters is as valid as the author's, and he or she will be happy to know that the rest of this impressive volume contains no more such speculations, but is a pragmatic, useful primer to those wishing to know more about this amazing technology, the camera phone.

Tom Killick
LensVector Inc.

Preface

The idea of creating this book was born, thanks to the fast penetration of miniature camera modules (MCMs) in our daily lives. These cameras have the ability to adjust their frame rate automatically with illumination change, focus at different distances, zoom, compensate for hand shake, and transform captured images. We can therefore refer to them as "smart" cameras. Beginning with webcams, these cameras have rapidly "invaded" mobile phones, tablets, portable computers, gaming, and interactive (smart) television systems, thus transforming our lives. Mobile video conferencing has become a reality. There is no doubt that many other uses for MCMs will become available in the coming years for industrial, medical, military, and security applications.

While the general concepts behind traditional digital still cameras (DSCs) are well known (and the corresponding literature is widely available), questions of the structure and performance tradeoffs of MCMs seem to be rather unknown to the general public (at the time of writing this text). At least this was my experience, when our team first started working on the development of liquid crystal autofocus (AF) and optical image stabilization (OIS) lenses for MCMs. Thus, the main goal of this book is first to provide information for students and engineers in their search for the best camera design with its component and performance tradeoffs. However, I hope that this book will also be appreciated by others who are curious about technology, to answer such "simple" questions as: How important is the number of megapixels in the mobile-phone camera we are going to buy? Is that the only factor to consider? How much should we value the AF or OIS features?

This book is structured in a way that provides key information about the MCMs and, to a lesser extent, some comparisons with traditional ones. The specific choices of MCM components (imposed by the size or the type of their application) in terms of optical design, image sensor, and functionalities are also presented. Several active functionalities (such as AF and OIS) are explained in their different implementations. Some future trends are also described.

The chapters are presented with minimal editorial changes to respect the authors' independence. I am happy to emphasize that valuable contributions are included by authors ranging from academics to field engineers. The book certainly does not cover all the technological aspects of MCMs, nor all the historical aspects of their development. However, I hope that this will be the first step, and that more contributions will become available in the near future.

I hope that the reader will find this book useful and enjoyable to read.

Editor

Tigran V. Galstian (or Galstyan) started his studies at Yerevan State University, Armenia. He received his MSc and PhD degrees (in quantum electronics) from the Special Department of Physics at Moscow Engineering Physics Institute, Moscow, Russia. He worked at different research institutions, such as the Institute of Applied Problems in Physics and Yerevan State University (Armenia), the Institut d'Optique (Orsay), and the Rennes I University (France), before joining (as professor) the Department of Physics, Engineering Physics and Optics at the Faculty of Sciences and Engineering of Université Laval, Québec City, Québec, Canada, where he founded two research laboratories. He also cofounded two high-tech companies (Photintech and LensVector).

His main research interests are new optoelectronic materials and their applications in the areas of optical information and biomedical optics, including behavioral biophysical systems. The core of many of his research and development projects is based on liquid crystal composite material systems, where light scattering and polarization are used as key tools. He has made up to 530 scientific contributions (publications, communications, and patents). He is president of the Canadian section of the International Optics Commission, associate editor of the *European Journal of Physics: Applied Physics*, and he is a member of the Center for Optics, Photonics, and Lasers, the Ordre des Ingénieurs du Québec, and the Société Française d'Optique.

Contributors

Dmitry Bakin
Heptagon Advanced Micro
    Optics
Mountain View, California

Bruno Berge
Parrot
Lyon, France

Andreas Brückner
Fraunhofer Institute for
    Applied Optics and Precision
    Engineering IOF
Jena, Germany

Peter P. Clark
LensVector Inc.
Maynard, Massachusetts

Tigran V. Galstian
Department of Physics,
    Engineering Physics and Optics
Laval University
Québec City, Québec, Canada

Tom Killick
LensVector Inc.
Sunnyvale, California

Luke Lu
LensVector
Shanghai, People's Republic
    of China

Junichi Nakamura
Japan Design Center
Aptina Japan, LLC
Tokyo, Japan

Eric Simon
Varioptic, Parrot
Lyon, France

Simon Thibault
Department of Physics,
    Engineering Physics and Optics
Laval University
Québec City, Québec, Canada

CHAPTER 1

Lens Design and Advanced Function for Mobile Cameras

Peter P. Clark

Contents

1.1 Introduction

The history of photography has been marked by developments that have dramatically changed the relationship of the photographer to the technology:

> *Imaging capability:* First there was only monochrome, then full color. First only still photography, then cinema and video. Light sensitivity has continuously improved. Most recently, digital imaging has been developed, giving the photographer powerful image-processing and editing capabilities.
>
> *Immediacy:* Until instant chemical photography became available in 1948, chemical photography required the photographer to wait for processing to see the image. In the 1990s, digital cameras with liquid crystal (LC) displays allowed immediate review of images. Cameras in mobile phones allow immediate electronic distribution of images.

Convenience: Early photographers used plates that needed to be prepared and developed near the camera, so when dry emulsions became available the darkroom did not have to be transported. Single-plate exposures with large cameras gave way to portable roll-film cameras that could make many images without being reloaded. Improvements to film and lenses shortened exposure time, allowing handheld photography of moving subjects and in low illumination. Electronics helped nonexperts, first with automatic exposure control and later with autofocus (AF) systems.

The size of consumer cameras became steadily smaller, from box cameras and folding roll-film cameras to 35 mm cameras, and then to even smaller consumer film formats such as Kodak's disc-camera system. Digital cameras continued the trend toward smaller size cameras, reducing the sensor size, and eliminating optical viewfinders and large strobe systems from most consumer cameras.

The miniature digital cameras that we see in many mobile phones and other products are an important step in that evolution. They allow the consumer the ability to record images at any time, with a device that is always at hand. The miniature camera modules (MCMs) are so small that their impact on the device size and portability is acceptable to the consumer, thus causing a revolution in the acquisition of still and video images. As with any digital camera, the images are instantly seen and using a mobile phone and the Internet, they may be instantly shared with anyone.

Furthermore, miniature digital cameras are being applied to a wide variety of applications in addition to conventional photography. Increasingly, automobiles are being fitted with small cameras for safety monitoring. Insurance companies employ cameras to monitor behavior, reducing accident rates and costs. Mobile-phone cameras are being adapted for technical applications, and even gigapixel "super cameras" are being assembled from large arrays of miniature camera sensors (Brady and Hagen 2009).

Table 1.1a and 1.1b show the comparative data for various types of cameras, configured for general photography. Table 1.1b extends Table 1.1a to include photographic performance information. This is not a complete survey of commercially available products. It is just a sample, intended to indicate trends and illustrate the changes due to camera scale.

TABLE 1.1 Comparison of Camera Formats

(a) Miniature Camera Modules, Digital Cameras, and Film Cameras[a]

Inch-Format	Horizontal (mm)	Vertical (mm)	Diagonal (mm)	Area (mm²)	Megapixels	Minimum Pixel (mm)	Maximum Pixel (mm)	Linear Scale (35 mm ref) (%)	Area Scale (35 mm ref) (%)	Typical Minimum f/number	EFL	Entrance Pupil Diameter
Miniature Camera Modules												
1/6	2.32	1.74	2.90	4.04	1.3–2	0.0014	0.0017	7	0.4	2	2.28	1.14
1/5	2.80	2.10	3.50	5.88	2–3	0.0014	0.0017	8	0.7	2	2.75	1.37
1/4	3.60	2.70	4.50	9.72	3–5	0.0014	0.0017	10	1.1	2.4	3.53	1.47
1/3	4.80	3.60	6.00	17.28	5–8	0.0014	0.0017	14	1.9	2.8	4.71	1.68
Digital Still Cameras												
1/2.3	6.08	4.56	7.60	27.72	12–16.6	0.0015	0.0022	18	3.1	2.8	6.0	2.1
1/2	6.40	4.80	8.00	30.72	16	0.0014	0.0014	18	3.4	2.4	6.3	2.6
1/1.7	7.44	5.58	9.30	41.52	10–12	0.0019	0.002	21	4.6	2	7.3	3.6
1	13.20	8.80	15.86	116.16	14.2	0.0029	0.0029	37	13	2	12.5	6.2
APS-C	23.60	15.80	28.40	372.88	12.2–24.7	0.0039	0.0055	66	43	2	22.3	11.1
FULL	36.00	24.00	43.27	864.00	18.1–24.7	0.0059	0.0069	100	100	1.4	34.0	24.3
Film Cameras												
Disc	11.0	8.0	13.6	88				31	10	2	10.7	5.3
APS-H	30.2	16.7	34.5	504				80	64	2	27.1	13.5
35 mm	36.0	24.0	43.3	864				100	100	1.4	34.0	24.3
6 × 6 cm	60.0	60.0	84.9	3600				196	385	2.8	66.6	23.8
4 × 5 in.	127.0	101.6	162.6	12903				376	1413	4.5	127.6	28.4

(b) Some Photographic Characteristics[b]

Inch-Format	Typical Minimum f/number	EFL	Relative Central Illumination (%)[c]	Relative Total Light Gathered (%)[c]	Resolving Capability			Depth of Field (Diopters)[d]	Closest Infinity Focus, mm (Hyp/2)[d]
					Diagonal (Airy_Disc_Diameter)	Diagonal (2*pixel)	Pixels (AD_Diameter)		
Miniature Camera Modules									
1/6	2	2.28	49	0.2	1,080	1,036	2.05	2.16	462
1/5	2	2.75	49	0.3	1,304	1,250	2.05	1.48	674
1/4	2.4	3.53	34	0.4	1,397	1,607	2.46	1.23	813
1/3	2.8	4.71	25	0.5	1,597	2,143	2.87	1.01	990
Digital Still Cameras									
1/2.3	2.8	6.0	25	0.8	2,023	2,533	2.68	0.798	1,254
1/2	2.4	6.3	34	1.2	2,484	2,857	2.46	0.650	1,539
1/1.7	2	7.3	49	2.4	3,465	2,447	1.51	0.465	2,149
1	2	12.5	49	6.6	5,911	2,735	0.99	0.273	3,662
APS-C	2	22.3	49	21.1	10,581	3,641	0.74	0.152	6,558
FULL	1.4	34.0	100	100.0	23,029	3,667	0.34	0.070	14,277
Film Cameras									
Disc	2	10.7	49	5.0	5,068			0.318	3,142
APS-H	2	27.1	49	28.6	12,858			0.125	7,971
35 mm	1.4	34.0	100	100.0	23,029			0.070	14,277
6 × 6 cm	2.8	66.6	25	104.2	22,582			0.071	14,000
4 × 5 in.	4.5	127.6	10	144.5	26,932			0.060	16,697

[a] Compared at the same diagonal field of view: 65° full (34 mm EFL for 35 mm format).
[b] "Total light gathered" is an estimate of how much light energy is collected to record the image. "Resolving capability" assumes a diffraction-limited lens.
[c] 35 mm = 100%.
[d] DOF based on the worst case of: 1- 2 pixel blur, 2- 1 Airy Disc blur, 3- Diagonal/1500 blur.

1.2 Key Optical Definitions

In this section, we introduce some key optical definitions, which are important to understand the operation and performance tradeoffs of miniature cameras. These are an incomplete introduction to the optics of imaging systems. Introductory optics texts, such as Smith (2007), should be consulted for more complete information.

Effective focal length (EFL): EFL is the separation of an equivalent ideal thin lens from the image it makes of an infinitely distant object (see Figure 1.1). EFL and subject distance determine the location and size of an image with respect to an ideal thin lens. The EFL may be positive (converging lens), infinite (e.g., a flat window), or negative (diverging), but it may not be zero (see optical power).

Optical power: The inverse of EFL, optical power is often expressed in diopters (inverse meters). It is useful to consider optical power because it easily handles the transition from positive to negative focal lengths. It is also a good way to describe the supplementary lenses sometimes used for focus correction (see Section 1.5.7).

Field of view (FOV): FOV is the extent of the captured image. The FOV may be described in object space or in image space. In image space, it is defined by the size of the sensor, either as x and y dimensions or as a diagonal dimension. In object space, it may be defined by the extent of the photographed object, but we often assume a distant object and measure it as an angle. One must be sure to understand if it is the "full" FOV or the "semi-FOV" (measured from the optical axis). FOVs of MCMs in mobile phones are currently 60°–75° full (corner to corner).

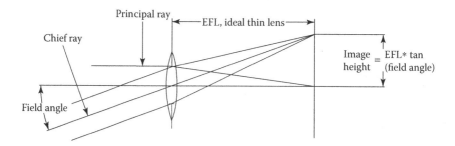

FIGURE 1.1 The relationship between the EFL of a thin lens, the field angle, and the image height. Infinite object distance.

Paraxial approximation: The paraxial approximation describes the behavior of a lens in the limit of small aperture size and FOV, greatly simplifying ray-tracing calculations. It tells us the ideal behavior of a lens system, ignoring geometric errors (aberrations). A real lens system that is designed and built to perform well is usefully characterized by its paraxial behavior.

Paraxial thin lens approximation: Any lens, or combination of lenses with a finite focal length can be substituted paraxially by a single thin lens. In this section, we will show some examples of lens behavior using the thin lens approximation. Figure 1.2, for example, illustrates a compound lens made up of three paraxial thin elements.

Optical axis: A lens system is usually rotationally symmetric about an optical axis. In most cameras, the optical axis is intended to intersect the sensor at its center point, and the sensor surface is normal to the optical axis.

Object space and *image space*: We will use the term *object space* to refer to the world outside the camera before light enters the lens system, and *image space* to mean after the light exits the optical system (the last lens surface). So, the sensor is in image space and the photographed subject is in object space.

Aperture stop: The aperture stop is the physical feature that limits the light passing through the lens, for the on-axis (central) field point (see Figure 1.2). The aperture stop is usually circular. In many cameras (but not in MCMs) its size is adjustable.

Entrance pupil and *entrance pupil diameter* (EPD) (see Figure 1.2): The entrance pupil is the image of the aperture stop in object space. We can see the entrance pupil if we look into the front of the camera lens to see the aperture stop. If the aperture stop is in front of the camera lens, the entrance pupil is identical to the aperture stop, but the aperture stop can be inside or behind the camera lens, so the location and size of the entrance pupil are not the same as the (physical) aperture stop.

Exit pupil (see Figure 1.2): Analogous to the entrance pupil, this is the image of the aperture stop in image space (in other words, as seen from the sensor).

f/number (fno): Also known as the relative aperture. Defined paraxially, this is the ratio:

$$\text{fno} = \frac{\text{EFL}}{\text{EPD}}.$$

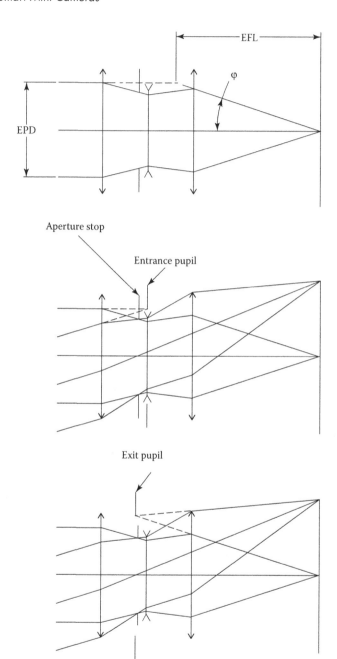

FIGURE 1.2 A triplet lens with three thin lens elements, an infinite object, with constructions from rays illustrating the EFL, aperture stop, and entrance and exit pupils.

More precisely,

$$\text{fno} = (2n \sin \varphi)^{-1},$$

where n is the refractive index in image space (usually air, $n = 1.0$) and φ is the angle of the real principal ray (see Figure 1.2). If $n = 1$, f/0.5 is the mathematically lowest possible fno, which is unrealizable in a camera; f/1.0 is a practical lower limit for camera lenses.

An image's light intensity is proportional to fno^{-2}. Traditional cameras have variable fno settings, often in factors of 2× exposure (f-stops): f/1.4, f/2.0, f/2.8, f/4, and so forth. Lower numerical fnos, then, provide more intensity in an image, and their Airy pattern is smaller (see hereafter), but their depth of focus is reduced and their geometrical aberrations may be larger. MCMs almost always have a single fixed fno. Currently, MCM fnos are typically between f/2.0 and f/2.8.

Rays: Lens design and analysis often model light energy propagation as geometrical "rays." The ray optics or "geometrical optics" model ignores the wave nature of light.

Chief ray: For a given field point, this is the ray that passes through the center of the entrance pupil and aperture stop (see Figure 1.1). The paraxial chief ray approximately tells us the scale of the image on the sensor, and the difference between the "real" chief ray and the paraxial chief ray at the sensor is a measure of distortion (see Section 1.4.4). The difference between chief rays of different colors tells us the lateral chromatic aberration.

Principal ray: This is the ray from the on-axis field point that passes through the edge of the entrance pupil and aperture stop (see Figure 1.1). Wherever the principal ray crosses the axis, there is an image of the object.

Point spread function (PSF): This is the distribution of light intensity in the image of an ideal point source. It might be the optical PSF (the optical image alone) or the recorded image PSF (optical plus sensor, etc.) (see Section 1.4.2).

Aberrations: These are geometric errors of the lens design or construction. Real imaging systems do not behave exactly like the paraxial ideal. If rays are traced exactly, we find that aberrations can keep them from forming perfect geometric images. Aberrations can be classified according to their functional dependencies. Aberrations may change with wavelength, and they often increase as the FOV and aperture increase. Most aberrations degrade image sharpness, but "distortion" aberrations cause errors of image size and shape without affecting sharpness. Defocus blur may be treated as a geometric aberration. The variation of defocus with wavelength is a form of "chromatic aberration." The correction of geometrical aberrations is the reason for much of the complexity of lens designs: multiple lens elements and aspheric surfaces are used to minimize ray errors in a lens design. Aberrations are also caused by lens fabrication errors, such as incorrect surface shapes and misaligned lens elements.

1.3 Construction of MCMs

1.3.1 Physical Construction

Currently, the mechanical construction of the lens of most MCMs is an assembly of injection-molded polymer parts. The sensor on a flexible printed circuit board is attached to a polymer "holder," which is equivalent to the body of a conventional camera. In a fixed-focus camera, that holder usually has a precisely threaded hole that receives the lens barrel. This thread is used for focus adjustment at assembly, then the barrel is fixed in the thread with an adhesive. The lens barrel contains the stack of lens elements and apertures, centered in the barrel by tight mechanical tolerances (<10 μm). The apertures in the barrel and the stack define the aperture stop of the system and block stray light paths. An infrared (IR)-cut filter (Section 1.3.2.2) may be mounted either in the lens barrel or in the holder.

The complexity of the lens design can vary. A very simple way to describe a lens design is with the number of elements: 1G2P means one glass element plus two plastic. Flat parts such as filters and windows are not counted. Most MCM lens designs at this time are 3P, 4P or 5P.

The prevention of contamination is an important consideration in the design and manufacture of MCMs. Very small specks of dust can cause visible defects in images if they settle close to the sensor.

An alternative method is "wafer-scale" construction. The wafer concept of construction of integrated circuits has been extended to the construction of lenses. A large array of replicated optical surfaces is constructed on each side of a wafer, then several wafers would be stacked together, perhaps even with a wafer of image sensors. Then, the entire assembly would be "diced" apart, producing hundreds of finished cameras. The technical challenges of this approach are significant, but it promises large cost reductions.

1.3.2 Electrical Construction and Photographic Performance

Electronic Shutter

MCMs generally have no mechanical shutter. The light integration time is started and ended electronically. Complementary metal-oxide-semiconductor (CMOS) sensors (see Chapter 2, this volume, for more details) use an electronic "rolling shutter" scheme. Exposure starts and ends in a raster scan, so different pixels are exposed at different times. If the integration time is short, there may be no time at which all of the pixels are collecting light. This is similar to the operation of the focal plane shutters in some film cameras (typically single lens reflex cameras), and it can cause distortion of images if the camera or the subject is moving during the exposure. It also makes it difficult to synchronize with electronic strobes. Most digital still cameras (DSCs) do incorporate a mechanical shutter, allowing better control of the balance between the flash and ambient exposure.

CMOS Image Sensor

The typical CMOS image sensors used in MCMs are an array of square detectors (pixels) with color filters. The arrangement is usually a "Bayer pattern" with a unit cell of four detectors:

green	red
blue	green

The "raw" image data from the colored pixels are interpolated to produce R, G, and B information for each of the Bayer sites (Fiete 2012). For example, a camera with one 4 megapixel (MP) sensor (2 MP of green detectors and 1 MP each of red and blue detectors) produces image files that have 4 MP × 3 colors. This is the most common pixel arrangement in larger digital cameras, as well. The need for circuitry and for the isolation of each detector from its neighbors requires that the light-sensitive area of a pixel be significantly less than 100% of its silicon area. This would cause a significant loss of collected light. To improve sensitivity, sensors use an array of microlenses, whose purpose is to make the detectors appear larger. As the pixels become smaller, the fill factor tends to become even lower, increasing the importance of the microlenses.

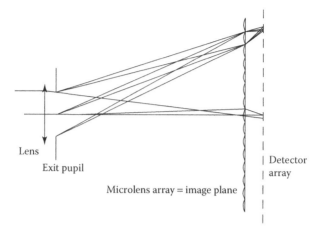

Lens

Exit pupil

Microlens array = image plane

Detector array

FIGURE 1.3 The microlens array images the exit pupil onto individual pixel detectors, improving light collection efficiency.

Figure 1.3 illustrates the microlens principle schematically. Note that each microlens is making an image of the exit pupil onto the sensitive area of the pixel. If that exit pupil image is smaller than the detector, none of the light that came to the microlens will miss the detector. With this arrangement, the image plane will be the microlens array, *not* the detector array. Notice also that the image sensor's response will be directional. Light exceeding a range of incidence angles might not be detected, or it may be incorrectly sensed by a neighboring detector. This is the source of the chief ray angle (CRA) requirement (Section 1.3.3.4). Since the microlenses serve to collect light, and not to form the image of a scene, they do not significantly affect the image focus or the sharpness, but if they are not made well enough, they could cause errors in exposure and color reproduction.

1.3.2.1 Pixel Size and Image Quality

There is much discussion about pixel size and image quality. Smaller pixels obviously allow higher spatial frequencies to be reproduced, and higher modulation transfer functions (MTFs, Section 1.4.2.2). However, smaller pixels collect fewer photons, so the signal/noise and dynamic range suffer. A good case can be made that the total sensor area correlates well with the subjective image quality, assuming that good design choices are made, but the desire for smaller cameras, lower costs, and higher pixel counts can lead to compromised image quality.

1.3.2.2 IR Sensitivity

Digital cameras with silicon sensors are intended to produce color images from visible light. The sensor's sensitivity, however, reaches its peak in the near IR (NIR). The color filters are not designed to block IR—in fact, all three color filters "leak" NIR radiation, so a separate IR-cut filter is included in the optical system. It usually has a sharp cutoff in the red, blocking light with wavelengths longer than about 650 nm.

1.3.3 Lens Design for MCMs

One could design a lens for these miniature cameras by simply scaling an existing design for a larger format. For example, a moderate wide-angle 33 mm EFL f/2.8 lens design for the 35 mm film format might be built at 1/10 scale, becoming 3.3 mm EFL f/2.8, appropriate for a 1/4″ format miniature camera. The image quality would be excellent. In practice, however, this is not what is done, largely for reasons of production cost and quantity. The cost must be less than one-tenth of the cost of the 35 mm solution, and production quantities can be millions of units per month, far higher than dedicated camera products (Bareau and Clark 2006).

Issues that affect the form of the camera lens design include:

1. *Materials*: To achieve production cost goals, the lens elements and structural parts must be injection-molded polymers. The 35 mm lens elements are made of a variety of optical glasses.
2. *Scale of the elements*: Manufacturing constraints can require the plastic parts to be thicker than the glass parts would be at the larger scale. This, and cost, limits the number of elements in the design. MCM objectives can be 3P, 4P, or 5P, while even fixed focal length 35 mm lenses have at least five glass elements and often more.
3. *Tolerances*: If the performance of the scaled 35 mm design was to be unchanged, the alignment tolerances of the lens elements would have to be scaled by 1/10×. This becomes very difficult, particularly with plastic parts.
4. *New requirements*: The miniature design may be required to have a shorter length, and CRAs need to be controlled to be lower than the field angles.

As an example, many conventional lens designs distribute optical power symmetrically about the aperture stop, automatically achieving correction of some aberrations, such as distortion and coma. The CRA requirement forces the aperture stop to be toward the object side of the lens assembly, eliminating that symmetry. To compensate, the design must become more complex. The designer turns to aspheric surfaces to achieve the required performance.

1.3.3.1 Aspheric Surfaces

Traditional glass lenses are almost always produced with spherical optical surfaces, as they are the easiest to make with a grinding/polishing process and they are simple to test (Malacara 2007). Aspheric surfaces can be applied to molded plastic lenses, because once the aspheric tooling is built, the production cost is no higher than that for spherical surfaces.

In the lens design process, aspheric surfaces are not a panacea. They do not allow perfect correction of image quality across the entire FOV, but they do add valuable design variables, similar to adding more lens elements. It is difficult to describe the function of individual aspheric surfaces in these designs. It is the combination of multiple aspherics within the design that enables simultaneous correction of ray aberrations and geometric distortion, along with controlling length, CRA, and relative illumination. Additionally, if not included in the design optimization, alignment tolerances can easily become so small that the design would be impossible to build correctly. The result is that nearly all the lens surfaces of these designs are aspheric and many are dramatically so, with departures from spherical surfaces that are easily seen with the naked eye (Figure 1.4).

1.3.3.2 Optical Materials

Chromatic aberrations are corrected in lens designs with the use of "crown" and "flint" materials, which have different degrees of refractive index change with wavelength (dispersion) (Smith 2007; Kingslake 1983; Kingslake and Johnson 2009). Optical glasses are available with a much wider variety of refractive index and dispersion than optical polymers.

Polymer optical materials differ from glass in some important respects. First, the range of refractive indices and dispersions available is limited. Their refractive index n and size change more rapidly with temperature; n decreases as the temperature increases. Also, some materials can absorb moisture, irreversibly changing their optical properties. Fortunately, a number of nonhygroscopic

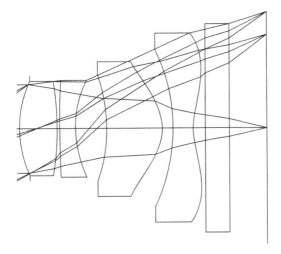

FIGURE 1.4 Cross section of an objective for a miniature camera module. The EFL of this 4P design is 3.4 mm, Total track (TT) = 4.3 mm, f/# = 2.3. The aspheric shapes are easily seen in the third (from *left*) and fourth elements. The actual edge details are not shown. The flat part after the lenses is the IR-cut filter (the vertical line on the *right* is the CMOS plane).

materials have been developed, essentially solving that problem. Finally, polymers' softness and low melting temperatures can affect the quality of vacuum-deposited coatings.

1.3.3.3 Physical Length

MCMs are usually required to be as short as possible, because the customers of mobile devices (that contain them) prefer them to be very thin. The camera lens accounts for most of the module thickness. "Total track" (TT) is the common measure: the TT is the distance from the first surface of the lens barrel to the sensor plane, usually measured at infinity focus.

The length of a lens design may be reduced by redistributing the optical power within the lens. We can illustrate this with a simple sketch of a lens. A thin single-element lens would be separated from the image of an infinite object by its EFL. Real camera lenses have multiple lens elements, so the optical power is distributed along the optical axis. Figure 1.5 shows a system with two thin positive lenses. When the front element has more positive optical power (shorter focal length) than the rear element, the TT becomes shorter even though the system EFL is constant.

This example is intended to be an illustration only. In practice, for MCM lenses, the ratio of TT to EFL is usually between 1.15×

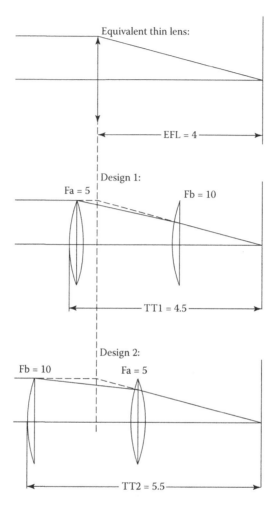

FIGURE 1.5 Redistributing the optical power in a lens design can change its length. In this thin lens example, the EFL is constant, and the TT/EFL changes from 1.125× to 1.375×.

and 1.30×. Forcing the design to be shorter would make it impossible to control geometrical aberrations. EFLs have sometimes become shorter to help reduce the TT.

1.3.3.4 Chief Ray Angle (CRA)

The CRA can be reduced by moving the aperture stop toward the object space, as shown in Figure 1.6, where the lens is represented as a single thin element. A special case would be reached if the aperture stop was at the front focal plane (separated from the thin lens by its EFL): the CRA would be zero for all fields; the system

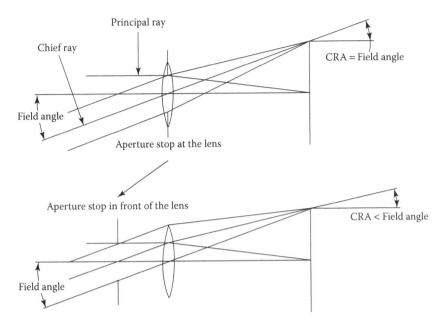

FIGURE 1.6 Chief ray angle is affected by the aperture stop location.

would be "telecentric." In practice, aspheric surfaces in the lens design can make the relationship of the CRA to the field angle very nonlinear. The CRA often reaches a maximum value at a large field angle, then it actually decreases from that to the corner of the image (see Figure 1.7).

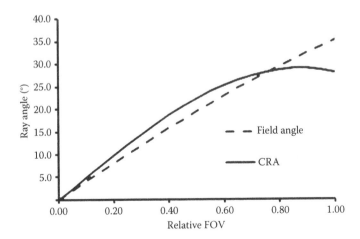

FIGURE 1.7 Typical chief ray angle plot for a miniature camera module. The maximum CRA is limited, and it decreases at the limit of the FOV. This is a 4P design for 1/4″ format, 70° full FOV.

1.3.3.5 Lens Design Process

The process of designing a lens for MCMs can be very complex, even with the aid of modern lens design software, and the designs become very different from any traditional glass lens forms. The aspheric surfaces require tracing many rays from many field points (at several wavelengths), and tolerance control can require simultaneously optimizing a number of configurations that include alignment errors. The design and construction of these highly precise optical systems at such low cost and high volume is an impressive and underappreciated accomplishment of modern optical design and manufacturing.

1.4 Basic Performance Aspects

1.4.1 Image Quality

Image quality can have many attributes, which are subjectively perceived and, often, objectively measured (Table 1.2).

1.4.2 Image Sharpness

Sharpness is a subjective quality of an image that tells us how well an object's contrast and details are preserved in the image. It does not consider exposure, color balance, noise, or other tonal errors, and it does not consider geometrical errors that affect only the shape of the image (distortion). Sharpness is often analyzed

TABLE 1.2   Attributes of Image Quality

Subjective Quality	Objective Measurement
Haziness, scatter	Veiling glare (VG) percentage (ISO 9358 1994)
Ghosts, flare	Photograph bright sources
Dark corners	Relative illumination, percentage (ISO 9039 2008)
Distortion (straight lines reproduced as curves, off-axis)	Optical distortion (percentage) or TV distortion (percentage)
Image sharpness	Limiting resolution (ISO 12233 1998) MTF
Image artifacts	Varies
Color inaccuracy	Photograph color chart
Tonal inaccuracy	Photograph gray step chart (ISO 14524 2009)

and measured using MTF and other metrics. Sharpness is never perfect; it can be changed by a number of effects:

1. The PSF is analogous to the "impulse response" of an analog electronic system. The PSF is desirably as small as possible, but in practice, it always has nonzero dimensions because of a number of effects:
   a. Diffraction.
      The wave nature of light causes light to be "scattered" (deviated from the original propagation's direction) from edges, a process called diffraction. In any lens system, light diffracts from the aperture stop, causing the image of a point source to be larger than geometrical optics would predict. If the aperture is round, the lens is perfect, and the incident light is uniform, the PSF takes the form of the Airy pattern, which, for a single wavelength, is a bright core surrounded by a series of rings (see Figure 1.8).
      The diameter of the first zero of the Airy pattern is

$$Da = 2.44\,\lambda\,\text{fno},$$

where λ is the wavelength (0.4–0.7 μm, visible light), and the entire pattern scales with the wavelength and the fno.
      We see that if the camera gets smaller and the image sensor becomes proportionately smaller, but the fno stays the same, then the Airy pattern becomes a larger fraction of the image. On the other hand, if the EPD remained constant, the fno would decrease proportionately and the Airy pattern/image size would be constant. It is not possible, however, to keep the entrance pupil constant

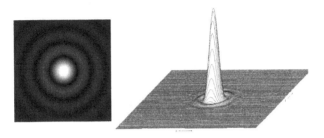

FIGURE 1.8 Image of a point source (PSF) by a perfect optical system with a round aperture stop: the Airy pattern.

as we go to very small cameras, because lens designs at extremely fast (low) fnos become unrealizable.

b. Geometrical aberrations cause light energy to spread, compared with a perfect point image (Figure 1.9).

c. Focus error. Similar to a geometrical aberration.

d. Image motion during exposure. This is caused by either camera motion or subject motion. It has a larger influence if the exposure time is long.

2. Pixel count. Sampling theory tells us that information with a spatial frequency (Section 1.4.2.2) higher than the Nyquist frequency = (sampling frequency)/2 cannot be reproduced.

3. Pixel size. The sensitive area of an individual pixel affects sharpness, just as the PSF of the lens does.

4. Image processing. Digital filtering techniques may be used to adjust image sharpness. Increasing the contrast (MTF) of a range of spatial frequencies is commonly done in digital photography. Digital sharpening causes some increase in noise and image artifacts, such as "ringing" at the edges, but a well-chosen degree of sharpening is usually beneficial.

One way to think about the imaging process is to consider the final image to be the convolution of the original object with the system's PSF. The PSF is, in this way, the "paintbrush" that creates the image.

The PSF and the optical transfer function (OTF) are alternative ways to describe the same thing. The OTF is the Fourier transform of the PSF. The PSF describes image quality in space; the OTF describes image quality in terms of spatial frequencies. It is often more convenient to use the OTF (and its modulus, MTF)

 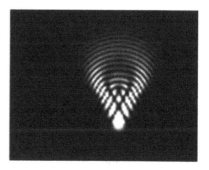

FIGURE 1.9 Example of a geometrical PSF of an aberrated star image from a grid of traced rays (*left*), and a similar PSF including the wave nature of light (*right*). The shape and size of the images compare, but the details differ significantly.

because in frequency space the OTFs may be combined by multiplying, which is simpler than convolving PSFs (see Goodman 2005).

1.4.2.1 Limiting Resolution

Limiting resolution is a measure of sharpness, determined (usually visually) by inspecting an image of a test target that has features that vary in size. The smallest feature that is judged to be recognizable is the limiting resolution. The optometric Snellen chart is a good example, as are the USAF 3-bar target (Figure 1.10) and some features in the ISO 12233 chart (ISO 12233 1998). Resolution testing can be useful, but it can be affected by the operator's judgment, and it emphasizes the smallest visible details, rather than the contrast of larger features in the image, which may correlate better with subjective image quality.

1.4.2.2 MTF

1.4.2.2.1 Spatial Frequency

A sinusoidal object varies in intensity from I_{min} to I_{max}. An image is information presented spatially in two dimensions, whereas most signals (audio, radio, etc.) are one dimensional (1D; time). Just as a 1D signal may be described by its temporal frequency content, any image may be built from a summation of sinusoidal intensity functions, over all azimuths and frequencies (Figure 1.11). The spatial frequency of a sinusoidal object of period, P, is 1/P, and its units are inverse length (cycles per millimeter, for example). A sinusoidal object is described completely by its frequency, azimuth, phase, and by the maximum and minimum intensities. Figure 1.12 shows the "modulation" of such an object. An imaging system can alter the phase and

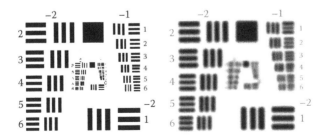

FIGURE 1.10 Limiting resolution. A portion of the USAF target (original, *left*, and blurred, *right*). The feature number (−1,3) at the right is the smallest with three bars visible in both orientations.

FIGURE 1.11 Sinusoidal objects. Spatial frequency examples: high frequency, mid frequency, and low frequency at a different azimuth angle.

modulation of the image of a sinusoidal object. This concept is useful because optical imaging systems behave as linear systems, and the reproduction of an image may be described by the system's frequency response (see Section 14.2.2.2).

1.4.2.2.2 OTF

The OTF is the spatial frequency response of an optical imaging system, including the amplitude and phase responses (Smith 2007;

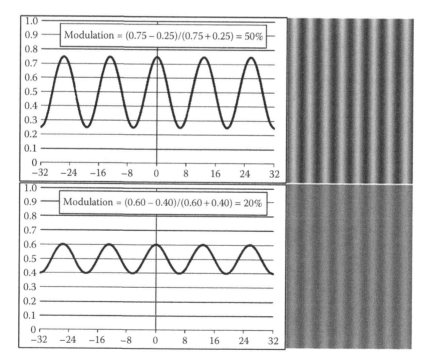

FIGURE 1.12 Two sinusoidal objects. 50% modulation and 20% modulation. Modulation = $(I_{max} + I_{min})/(I_{max} - I_{min})$.

Goodman 2005). The MTF is the modulus of the OTF. It refers only to the amplitude component, ignoring the phase errors. An ideal imaging system would have 100% MTF at all frequencies, but even an ideal lens (geometrically perfect) reduces the MTF because of diffraction caused by its aperture. If a lens is not geometrically perfect, it may suffer from aberrations, or defocus, and the MTF is further reduced. Figure 1.13 is an example of MTF (spatial frequency) curves for a perfect lens, and for the same system with varying degrees of aberrations. MTF at zero spatial frequency is always unity. In Figure 1.12, the photograph on the right could be the image of the object on the left formed by a system with 40% MTF.

1.4.2.2.3 Digital Sharpening

While the optical system and the detectors inevitably reduce MTF, it is possible to improve sharpness in an image by postprocessing the information. A digital filtering algorithm may be used to amplify the image signal as a function of spatial frequency. Noise will be amplified, as well as the signal, and if the gain is too high there will be visible artifacts, such as "ringing" at the edges. Nevertheless, digital imaging systems almost always incorporate some degree of digital sharpening. Interestingly, silver halide emulsions would often have some "sharpening" built

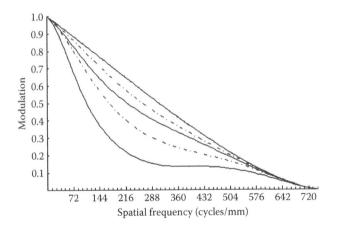

FIGURE 1.13 MTF versus the spatial frequency for an optical system. On-axis it is geometrically perfect (the highest curve in this figure), but off-axis, geometric aberrations reduce the MTF. Notice that MTF(0) is always 1.0 (100%). The off-axis field point curves apply to 50% and 100% of the FOV for this system. There are two curves for each because the MTF is different in the tangential and the radial directions. MTF is zero above 722 c/mm, which is the diffraction cutoff frequency for this lens.

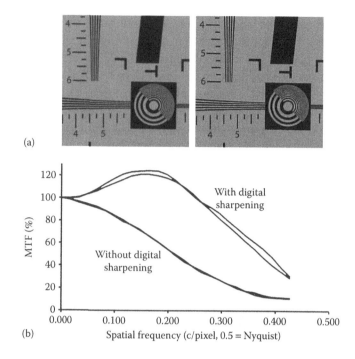

FIGURE 1.14 (a) Image of a test target, without (*left*) and with (*right*) digital sharpening. With sharpening, exaggeration of the edges is visible. (b) Measured MTFs from the two images in (a). MTFs greater than 100% are not possible in the unprocessed optical image.

in, due to the adjacency effect—film MTFs could be greater than unity over a range of spatial frequencies (Fiete 2012). See Figure 1.14a and 1.14b for an example of measured MTFs with and without sharpening.

1.4.3 Relative Illumination

Relative illumination is the ratio of image intensity off-axis to image intensity on-axis. Generally, illumination decreases as the field angle increases. It is influenced by the geometry of the relationship between the aperture stop and the flat image plane. If the camera lens is represented as a perfect thin lens with the stop in contact, CRA = field angle, and illumination falls off in proportion to cos(field angle)^4.

In practice, relative illumination is also influenced by other factors. For example, if the CRA can be reduced, as in Figure 1.15, relative illumination is increased. Figure 1.15 shows the relative illumination function for the classic cos^4 case and for a design

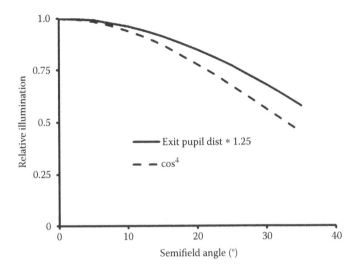

FIGURE 1.15 Relative illumination can be changed by changing the exit pupil distance.

where the CRA is lowered by increasing the exit pupil distance. Also, the lens design can cause distortion of the shape and size of the entrance pupil, changing the relative illumination—and affecting other aberrations too, particularly distortion.

Relative illumination in MCM lenses is frequently specified as not less than 40%–50% in the corners. The dark corners in the optical image (referred to as "lens shading") are at least partly corrected by digital processing, at the cost of increased noise.

1.4.4 Dynamic Range

Loosely speaking, the dynamic range of an image is the ratio of the light intensity between the lightest and darkest distinguishable levels. In digital cameras, it is strongly dependent upon pixel size, because larger pixels can detect more photons before reaching saturation. Cameras with a limited dynamic range can lose information at the light and dark extremes of a scene, giving an unpleasant flatness, particularly to highlights (ISO 14524 2009).

1.4.5 Distortion

Optical distortion is an aberration that causes a geometric displacement of the image, and not a loss of sharpness. Distortion can be a property of the lens design, and it may be thought of as a variation of the image magnification with the field. In lens design, the convention is to calculate distortion as a percentage (Smith 2007; ISO 12233 1998):

$$\text{Distortion} = \frac{\text{Real chief ray height}}{\text{Paraxial chief ray height}} - 1.$$

If distortion is positive, it is often referred to as "pincushion" distortion. If it is negative, it is called "barrel" distortion. Sometimes, distortion can be more complicated—varying from positive to negative as the field changes.

An alternative measure of distortion is called "TV distortion." In fact, there are several suggested methods for measuring TV distortion. For example: If we make a photograph of a grid that nearly matches the picture format, TV distortion is the following:

$$\text{TV distortion} = \frac{\text{Average of left and right vertical edges of the image}}{\text{Vertical height of the image at the center}} - 1.$$

TV distortion is easy to measure from photographs; it measures the effect upon a straight line that is near the edge of the photograph, which is a source of user objections.

The specifications for distortion in MCM lenses are typically <1%–2% optical, <1%–1.5% TV. Usually, higher distortion values are accepted if the FOV is larger.

Figure 1.16 shows some examples of distortion (they are more severe than usual in an MCM for easier visualization). It is not unusual to see a higher order, as shown in Figure 1.16d, in an MCM because of the complex aspheric surfaces in the lens.

1.4.6 Sampling and Aliasing

Digital imaging systems employ regular arrays of small detectors, to "sample" the light image (Fiete 2012; Goodman 2005). The goal is usually to reconstruct the image accurately, but the sampling process imposes a limitation. The Nyquist–Shannon sampling theorem tells us that if the light image being sampled contains no information at spatial frequencies higher than half of the sampling frequency, called the Nyquist frequency, then the image may be perfectly reconstructed from the sampled data. Higher-frequency content in the light image, however, cannot be distinguished from low-frequency information, so it will be inaccurately reproduced, producing low-frequency artifacts in the reconstructed image. This process is called "aliasing."

To illustrate, a digital image sensor with pixels spaced at 1.4 microns will have a sampling frequency of $1/0.0014$ mm$^{-1} =$ 714 cycles/mm. Its Nyquist frequency is 357 cycles/mm.

Aliasing may be eliminated by low-pass filtering the light image before detection, but it is not possible to do this perfectly without losing image sharpness. Some cameras (e.g., DSCs) use birefringent antialiasing filters that create multiple copies of the PSF, usually in a square array of four points. Some have used aspheric elements designed to increase blur. Small digital cameras use a single image sensor with color filters, such as a Bayer pattern array. In such systems, aliasing becomes color dependent,

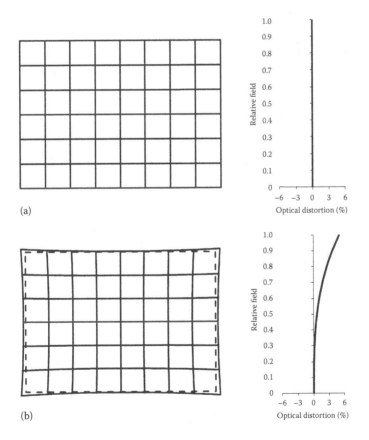

FIGURE 1.16 Distortion of various orders. (a) Zero distortion. Straight lines in the object are straight in the image. (b) Pincushion distortion (+5% optical, +3.9% TV). (c) Barrel distortion (−5% optical, −4.0% TV). (d) Higher order (−3.6% max optical, 0% TV).

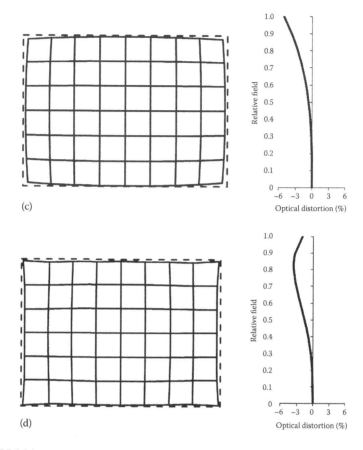

FIGURE 1.16 (Continued)

because the red, green, and blue samplings are different. Some image-processing methods are used to reduce the subjective effects of color aliasing. Systems with multiple sensors, or with unusual sensors that can detect all three colors in the same pixel, can have a significantly better aliasing performance.

1.4.6.1 Aliasing in MCMs

MCMs almost always use Bayer array sensors, and they do not usually have antialiasing filters. MCMs' image sharpness is closer to "optics-limited" than larger cameras, because the pixels are so small. If we compare the Airy disc diameter to the pixel, a full-frame DSC (35 mm format) is likely to have an f/2.0 lens, and 6 micron pixels. The MCM lens will have a slower fno (larger diffraction PSF), and 1.4 micron pixels. We can create MTF plots for both cases (Figure 1.17). The dashed line is the system MTF

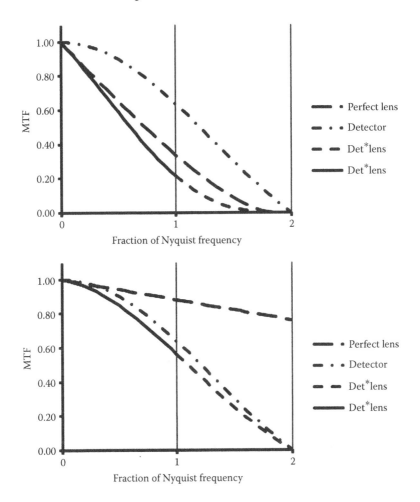

FIGURE 1.17 Aliasing comparison, 1/4″ MCM (*top*), full-frame DSC (*bottom*). The area under the dashed MTF curve is the potential for aliasing if the object has a frequency content higher than Nyquist.

at frequencies above the Nyquist frequency. The integral of the system MTF above the Nyquist frequency tells us how severe the aliasing will be, assuming that the object has content at frequencies above Nyquist. It is apparent that the DSC has the greater need for an antialiasing filter.

1.4.7 Stray Light

Stray light refers to light that takes an unintended path to get to the image. In this context, it does not include the sharpness-decreasing effects of lens defocus, aberrations, and diffraction. Instead, there are two broad categories: flare and veiling glare.

1.4.7.1 Flare, Ghosts

Flare, often called ghosts, can be difficult to diagnose, because there are many possible sources. Light from bright sources in front of the camera (maybe within the FOV, maybe not) can end up on the sensor where it does not belong. There may be paths around the lenses that were unintended, or there may be reflections from a structure within the camera, or reflections from the lens surfaces. There is no consistent objective measure of flare, because its appearance can vary widely. It is often tested by photographing a bright source over a wide range of camera angles. Flare can be minimized by the use of antireflection coatings on lens surfaces, by control of lens surface shapes to avoid focused reflected images on the sensor, and by the use of internal baffles (Figure 1.18).

1.4.7.2 Veiling Glare

VG is caused by scattering of light (ISO 9358 1994). It is distinct from flare or ghosts because it has little structure in the image. VG is the overall haziness in a photograph, increasing the black level. Defined as the fraction of the average exposure of a photograph that is scattered (uniformly) across the photograph, it is usually caused by scatter from hazy, damaged, or dirty optical components. The most common measurement method is to photograph a small black object in a white field. Frequently, an integrating sphere is used to create the hemispherical white field that illuminates the lens system (Figure 1.19). VG is often under 5%. VG may be reduced by avoiding out-of-field light—the photographer's lens hood does this. The effect of VG may be somewhat reduced by image processing to reset the black level of the image.

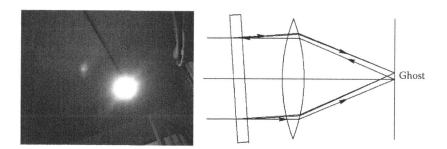

FIGURE 1.18 Ghost image example. Light from the lamp's image is reflected from the sensor, and then re-reflected by the two surfaces of the window. A ghost image is seen to the left of the lamp. 5 MP mobile-phone camera.

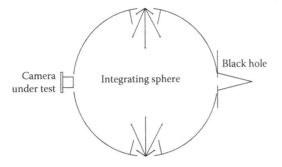

FIGURE 1.19 Schematic of the veiling glare test.

1.5 Focusing Issues in MCMs

1.5.1 Simple Thin Lens

To a paraxial approximation, an imaging system creates a 3D "image" in image space of the object space volume. Every plane normal to the optical axis in the object space is imaged in a plane in image space. The familiar, simple thin lens equation describes the focus distance relationships:

$$\frac{1}{\text{EFL}} = \frac{1}{z} + \frac{1}{z'},$$

where z is the object plane to the lens distance, and z' is the lens to the image plane distance (see Figure 1.20). The algebraically equivalent form of this equation, in which distance is measured from the front-focal and back-focal planes of the simple lens, is

$$\text{EFL}^2 = s\,s',$$

where $s = z - \text{EFL}$ and $s' = z' - \text{EFL}$.

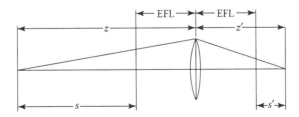

FIGURE 1.20 Variables in the thin lens equations.

So, we see that closer objects are imaged further behind the lens, and the focus shift from an object at infinity is

$$s' = \frac{EFL^2}{s}.$$

As long as $s \gg EFL$, s may be considered to be equal to the object distance, which is almost always the case for miniature cameras, whose EFLs are very short.

1.5.2 Depth of Field and Depth of Focus

The depth of field and the depth of focus are the range of acceptable focus, measured in object space (...field) or image space (...focus). The photographer is concerned with the depth of field.

Determining the depth of field requires us to have an upper limit for acceptable blur in an image. Blur due to defocus may be measured geometrically by tracing a principal ray. The blur diameter, B, is twice the absolute value of the ray intercept on the image plane (see Figure 1.21).

1.5.3 Fixed-Focus MCMs

Many miniature cameras have no focus adjustment capability, so they have a limited range of subject distances with acceptable sharpness. Best focus is usually set so that sharpness is just acceptable for infinite objects, and best focus is found at a closer distance, referred to as the "hyperfocal" distance, H. The blur at $H/2$ is approximately the same as the blur at infinity, so the camera will have acceptable sharpness from infinity to $H/2$. If the limit of acceptable blur is some multiple, k, of the pixel dimension, P, then for a lens set to the hyperfocal distance, fno = $S'/(kP)$:

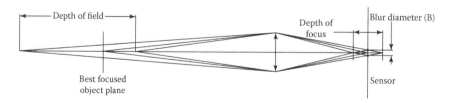

FIGURE 1.21 Depth of field and depth of focus. Depth of field is not usually symmetric about the best focus distance.

$$\text{fno}\, k\, P = \frac{\text{EFL}^2}{H}$$

and

$$H = \frac{\text{EFL}^2}{\left(\text{fno}\, k\, P\right)}.$$

Example: EFL = 3 mm, f/2.8, $P = 0.0014$ mm, $k = 2$ pixels:

$$H = 9/(2.8 \times 2 \times 0.0014) = 1148 \text{ mm and}$$
$$S' = \text{fno}\, k\, P = 0.008 \text{ mm}.$$

So, this fixed-focus camera will have acceptable sharpness for objects from infinity to 574 mm.

1.5.4 Adjustable Focus MCMs

Adjustable focus MCMs are often required to focus from infinity (zero diopters) to 100 mm (10 diopters), so fixed focus may not be adequate. If we want to build a camera that can photograph closer objects sharply, it needs to have some means of focus adjustment. One way to accomplish this is by allowing the distance between the lens and the sensor to be adjustable, so the best focus plane can be closer to the sensor. The motion required to focus from infinity to the closest distance, S_{min}, is approximately proportional to $1/S_{\text{min}}$:

$$S' = \frac{\text{EFL}^2}{S_{\text{min}}}.$$

If EFL = 3 mm again, and $S_{\text{min}} = 100$ mm, we need 0.09 mm of motion, and another question may be asked: If the focusing could be done in discrete steps, rather than continuously, what is the smallest number of steps that would be required? The following calculation helps to answer:

$$\text{No. of steps} = \frac{S'_{\text{total}}}{\left(2\, S'_{\text{hyperfocal}}\right)} = \frac{0.09 \text{ mm}}{\left(2 \times 0.008 \text{ mm}\right)} = 5.7 \text{ focus steps}.$$

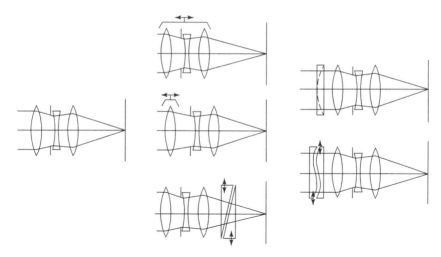

FIGURE 1.22 Means of focusing MCMs. (*Left*) Fixed focus. (*Center*) Axial distance adjustment. (*Top* to *bottom*: Unit focusing, front element, and wedge prism.) (*Right*) Adjust optical power. (*Top*: Liquid, LC, or deformable. *Bottom*: Alvarez lens.)

The strategy of moving the entire lens, just described, is common. It is referred to as "unit focusing." Some systems move only part of the lens ("partial lens focusing")—perhaps just the front element. Partial lens focusing is attractive because the mass to move is usually much smaller, and if the EFL of the moving part is shorter than the EFL of the entire lens, the length of the motion can be reduced. The drawback, however, is tighter mechanical tolerances for decenter and tilt, and also the possibility of a larger aberration variation with focus change. The partial lens focusing method requires a special lens design and probably an advanced lens suspension system.

Other methods are possible. Figure 1.22 illustrates some focusing strategy alternatives: We could move only part of the lens, or we could effectively adjust the lens to sensor distance with a wedge prism set. Also, we could focus by changing the optical power of the lens system. Some of these methods are described in the following sections.

1.5.5 AF Systems

Before AF systems were first developed in the 1970s, photographers needed to adjust the camera focus. Originally, they would replace the film or plate with a ground glass screen, allowing direct inspection of the image. Since then, some cameras have required the photographer to estimate the distance to the subject and set a distance scale on the camera, some have used optical range finders to measure the distance and couple it to the lens motion, and "reflex" cameras have allowed ground glass focusing without moving the film.

AF systems have employed a number of means to judge image focus. Some have used active systems (IR or ultrasound) to measure the subject distance. Others have automated the optical range finder, looking for a correlation of two optical images. In digital cameras, including miniature digital cameras, the simplest method is to continuously collect image data, and compute a "focus score" (FS; more properly, a sharpness score). The FS is usually determined from a subsection of the image, the region of interest (ROI). The ROI is usually in the center, but it can be set by the user in some devices. The AF algorithm changes the focus setting, and then readjusts the focus, trying to maximize the FS. Different focusing technologies may require different algorithms, and the AF algorithm for a still photograph is likely to be different from the continuous AF algorithm that is required for video recording.

One of the goals of an AF system design is to minimize the response time. For still photography, the AF process starts when the shutter button is pressed. The time that the AF algorithm takes to find and set the best focus adds to the "shutter lag." Naturally, camera designers strive to minimize the shutter lag; in many camera systems, it is on the order of 1 s. Some cameras achieve much shorter times. The response time in video systems should be fast enough for the system to track moving objects and to respond quickly to sudden scene changes.

1.5.6 Mechanical Systems

Currently, the focus adjustment of MCMs is most commonly accomplished by an electromechanical arrangement called a voice coil motor (VCM; discussed in Chapter 3, this volume). As the name suggests, it is a drive very much like the one in a loudspeaker, with a fixed permanent magnet and the lens mounted on a suspension arrangement with wire coils that can be energized to move the lens. This arrangement can work very well, but it requires careful design and manufacture of the suspension. It is important to keep the lens from tilting as it moves, and care must be taken to overcome gravity effects and hysteresis.

Other mechanisms of discrete movements of MCM elements (to achieve AF) are known, such as microelectromechanical systems (MEMS), piezo elements, and shape memory alloys. However, to date, none of these has been widely applied in AF MCMs.

1.5.7 Electrically Variable Lenses

Much effort has been invested in developing tunable lenses—lens elements whose optical power can be varied by electrical signal. This is a way to focus the camera without any conventionally moving parts. Currently, three approaches to this are known to this author: electrowetting liquid lenses, deformable lenses, and LC lenses.

Adding a tunable optical element to a fixed camera lens has great potential for AF without moving parts. The tunable element should be located reasonably close to the aperture stop. If it is not near the stop, its diameter must grow, because of the camera's FOV. Tunable lenses have a limitation on the "optical path difference" (OPD) between their center and edge. If we wish to achieve a camera focus variation of, say, 0 to +P diopters (the subject distance from infinity to $1/P$ m), in a camera whose entrance pupil diameter is EPD (mm), the optical path difference across the aperture stop must be

$$\text{OPD}_{\text{min}} \sim= \frac{\left(\text{EPD}/2\right)^2 P}{2000}\,\left(\text{mm}\right).$$

If the tunable lens is assumed to be thin, and we consider the paraxial principal ray and chief ray heights on the tunable lens (Y_{prin} and Y_{chief}, respectively), then we can calculate the OPD required for the tunable lens to achieve the focus range:

$$\frac{\text{OPD}}{\text{OPD}_{\text{min}}} \sim= \frac{\left(abs\left(Y_{\text{prin}}\right) \pm abs\left(Y_{\text{chief}}\right)\right)^2}{Y_{\text{prin}}^2}.$$

We can look at a paraxial example: If the camera lens is a thin lens (see Figure 1.23), 3 mm EFL, f/2.4, and 65° full FOV.

Table 1.3 shows the OPD/OPD$_{\text{min}}$ required for 100 mm close focus distance.

If the tunable lens is placed in the image plane, it loses its focusing ability completely. This paraxial exercise is approximate, but it illustrates the importance of the location of the tunable lens in the optical system.

It should be understood that the addition of a tunable lens of any type affects the aberration balance of the camera system. Usually, sharpness will be optimized for the more distant end of the focus range, because (a) more distant photographs than close-ups are taken; and (b) while many distant scene subjects are at "infinity"

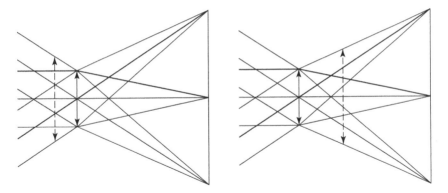

FIGURE 1.23 (a) A tunable lens (*dashed*) is located ahead of the entrance pupil (*left*) and (b) after the exit pupil (*right*). Increasing the distance from the pupil increases the OPD required to focus (Section 1.5.7).

across the entire FOV, close subject focus can vary widely with the field, so slightly less corner sharpness can be acceptable.

1.5.7.1 Electrowetting Liquid Lenses

As discussed in Chapter 5, this volume, electrowetting liquid lenses have the form of a cell with two flat windows that contain two immiscible liquids (oil and water). The oil and water arrange themselves like a doublet lens, with a curved interface between them. The geometry of the cell is such that the curvature of that interface changes as a voltage is applied to the part. The oil has a higher refractive index than the water, so the optical power changes. The material choice is critical. For example, it is important that the densities match closely, to avoid gravitational and thermal issues. Quite large optical power variation is achievable.

1.5.7.2 Deformable Lenses

Generally, deformable lenses are made with a liquid or polymer core material that is flexible. The edges are mechanically squeezed

TABLE 1.3 OPD/OPD$_{min}$ Required for 100 mm Close Focus Distance

Tunable Lens to Pupil (mm)	Case A (Figure 1.23a)	Case B (Figure 1.23b)
0	1.0×	1.0×
0.5	2.3×	2.6×
1.0	4.1×	6.4×
1.5	6.4×	16.5×

or stretched (e.g., by piezoelectric actuators), changing the curvature of the lens surface.

1.5.7.3 LC Lenses

As discussed in Chapter 6, this volume, LC lenses use the birefringence of LC materials to create a gradient-index optical element. Applying an electric field to an LC will rotate its molecules, and if the electric field is controlled spatially, the LC material will behave like a tunable lens.

1.5.7.4 Other Optical Focusing: Lateral Motion

Focusing in discrete steps may be accomplished by moving separate fixed focal length lenses in front of the camera. The extra lenses can be quite weak, so mechanical tolerances would be reasonable, but the volume required for the lenses and the actuator has kept this idea from being implemented in mobile devices.

It is also possible to achieve continuous focusing by using complementary unsymmetric aspheric elements (the "Alvarez" lens is an example); one may be fixed and the other may translate linearly or rotationally. These have been used in a small number of larger format cameras, but they are not yet in commercial miniature cameras (Plummer et al. 1999; Zhou et al. 2012).

1.5.8 Alternative Approaches

1.5.8.1 Software Focusing and Extended Depth of Field

As discussed in Chapter 4, this volume, several proprietary methods have been developed that eliminate all physical focusing, using special optical designs and sophisticated image processing to provide sharp images over a wide range of object distances. We will not describe them in detail here, but two general approaches have been seen:

1. *Extended depth of field (EDOF)*: A special optical design is produced with aberrations that reduce the variation of the PSF with focus. The recorded image is unsharp, but processing can then restore sharpness by deconvolving that constant aberration. All distances within the range of capability are acceptably sharp.
2. *Chromatic focusing*: The camera lens is designed with an intentionally uncorrected chromatic aberration. Chrominance and luminance are separated for processing, and the sharpest color information is used in a higher proportion in the luminance channel.

These descriptions are greatly oversimplified, and are intended only as descriptions of the principles.

EDOF systems have obvious strengths and weaknesses:

Strengths:
- No moving parts or physically tunable optics.
- No "shutter lag" due to focusing.
- AF decision errors are eliminated.

Weaknesses:
- Special optical designs may be difficult to design, manufacture, and test.
- Complex image processing in the device.

Image artifacts:
- Noise may be amplified. Algorithms can result in some spatially varying image artifacts.
- Somewhat unnatural results—photographers may prefer less depth of field.

1.5.8.2 Light Field Photography

Recently, some unconventional advanced photography systems have been developed. One of these is "light field photography" (Ng 2006). This approach enables EDOF, or refocusing after acquisition, using a special sensor arrangement. A simple statement of the principle is that the entrance pupil is spatially divided into a number of smaller segments, and image data are collected for each of those subpupils (see Figure 1.24). We might think of each subpupil as creating a long depth-of-field image (high fno). Light from the separate subpupil images is combined, spatially shifted, to give us a post-focused image. The post-processing options are flexible, allowing post-focusing or extended focusing. Another way to look at it is to realize that the light field sensor records angular information as well as spatial information. Much more pixel data are recorded for the construction of a finished photograph than a conventional digital camera photograph and much more processing must be done to display it.

1.5.8.3 Array Cameras

It has been proposed to create a thinner camera system by building an array of smaller, lower-resolution cameras to replace a

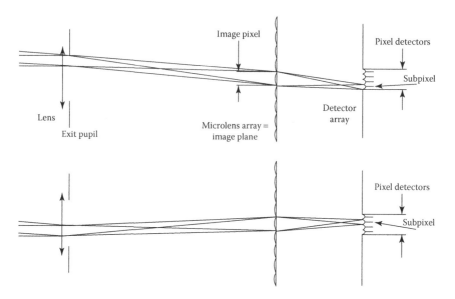

FIGURE 1.24 Light field photography principle. Four subdetectors can be seen behind each microlens. Each subpixel sees light only from a portion of the exit pupil. Two cases are shown, seeing the same pixel of object space, but using different parts of the lens pupil.

single, high-resolution device. While many arrangements are possible, an example would be to replace one N-MP camera with four $N/4$-MP cameras (Figure 1.25). The pixel dimension would remain the same. Each "quarter camera" might have the same FOV as the full camera, so the lenses could be 1/2× scaled versions of the original—half as long, with the same fno. The FOV, pupil area, and total detector area would be unchanged, so the system's total light collection is unchanged. Each of the quarter cameras could have a single color filter, instead of a filter array on the sensors. If two cameras were green, and one each red and blue, they might be arranged so that the pixel centers formed the familiar Bayer pattern when projected into object space. The big difference, compared to the full camera, is that the pixels will be twice as large in object space, overlapping each other. This would reduce the MTF somewhat, because the size of the sensitive area of the pixels affects the object space PSF. It also reduces aliasing. Other arrangements of array cameras, plus sophisticated image processing, might give us improved imaging in the future: maybe a wide dynamic range, post-acquisition focusing, etc.

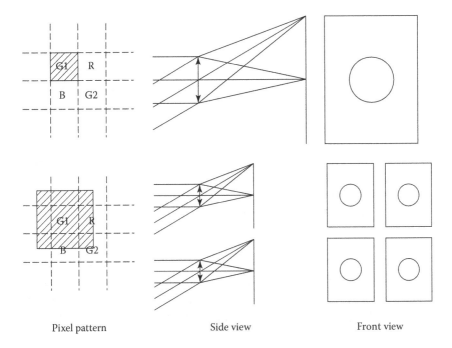

| Pixel pattern | Side view | Front view |

FIGURE 1.25 Four half-scale cameras in an array compared with the original. (*Left*) The pixel patterns seen to scale in object space. Each detector is two times larger in the array camera, but the Bayer pattern can be retained if the cameras are aligned.

1.6 Adapting Mobile Device Cameras to Special Applications

Miniature cameras are almost universally available in mobile phones, tablets, and laptop computers. Various "apps" may be created that operate the camera and process the images, so it is attractive to use them as a means for data collection that is widely available, portable, low cost, "smart," and connected. External hardware can be built to provide optical input, for microscopy, spectroscopy (Gallegos et al. 2013), and other types of optical measurements. Many applications can be imagined in commerce, science, medicine, etc.

When designing instruments to interface with miniature cameras, we need to consider their characteristics:

1. *Input characteristics*:
   a. Entrance pupil: All of the light collected must get into the entrance pupil. It is essential to know its size and

location. Fortunately, it is not difficult to measure from outside the camera with a good measuring microscope.

b. FOV: The size of the recorded object space, usually measured angularly, is important to know. Assuming that distortion is well-corrected by the camera lens, height in the image will be proportional to tan(field angle).

c. Angular size of pixels: Again, the pixel pattern divides tan(FOV) evenly. The FOV and the pixel size are also easy to determine by photographing a known target at a known distance.

2. *Spectral characteristics*:

a. The IR-cut filter will block light beyond about 650 nm.

b. The red/green/blue (RGB) color filter array is designed to provide pleasing color pictures, but for technical use, the images may need to be calibrated for spectral sensitivity variation.

3. The normal output of the camera will have been processed to provide visually pleasing photographs.

a. The preprocessing "raw" image may be desired for technical applications, but it might not be available to the app.

b. Processed images will probably have undergone image processing:

    i. Gamma correction. Pixel values are given a nonlinear relation to light exposure.

    ii. Lens shading correction. Corners are lightened.

    iii. Color interpolation, providing RGB information for every pixel location.

    iv. Automatic white balance correction. The software attempts to optimize color balance based upon image data.

    v. Sharpening. MTF is boosted at some spatial frequencies.

    vi. Noise reduction.

    vii. Compression (.jpg files are typically compressed, with consequent loss of information).

4. *Camera operation*:

a. The camera may have an AF system that adjusts the lens focus.

b. Autoexposure. Integration time and gain may be automatically adjusted.

c. Image stabilization. Not very common yet, but cameras may have a system that corrects for camera shake.

Which issues are important should be determined by the needs of the application, of course. The device may allow camera operation conditions to be set by the software. It may be necessary to disable some automatic functions, for example.

1.7 Advanced Capabilities of MCMs

1.7.1 Image Stabilization Systems

Image stabilization systems, described in Chapter 7, this volume, are intended to reduce the effect of camera motion (shake) on the recorded images. In still photography, camera motion during an exposure can result in loss of sharpness. Correcting camera shake in still photography will allow the photographer to make longer handheld exposures, which is a great benefit in low-light situations. With video, camera motion can cause instability of the scene, which is not dependent upon scene brightness. Like any free body, a camera can move in six axes, three translations and three rotations. If the object distance is large, pure translations have very little effect upon the image, so x, y, and z motions are generally not corrected. All three angular tilt modes, however, can cause loss of sharpness or disturbing image instability. Small tilts about the x- and y-axis (pitch and yaw) cause lateral shifts of the image, and rotation about the optical axis (z-axis—roll) causes rotation of the image.

1.7.1.1 Optical Image Stabilization

If the mobile device has a gyroscopic sensor, it can be used to acquire tilt information in real time. Tilt data are used to move some components of the imaging system to cancel the image motion. We can think about this problem as a control system that responds to inputs over a range of temporal frequencies. There must be a low-frequency roll-off, since the camera should not try to correct slow changes, and there will be a high-frequency roll-off because we do not expect very high-frequency shaking to have significant amplitude. Within the frequency range, the maximum amplitudes required to correct will vary, based upon the human camera shake. The phase delay of the correction must be small, or the correction will be ineffective. The correcting motion may be accomplished in several ways: (a) tilting of the entire camera module; (b) lateral movement of the image sensor, or of the lens or part of the lens; or (c) using a tunable lens technology to create an adjustable prism. Usually, roll is not corrected in optical image stabilization (OIS) systems—moving lenses or thin prisms cannot rotate the image.

1.7.1.2 Software Image Stabilization

Video may be post-processed to stabilize the images. It is more difficult to improve the sharpness of still images after the fact, but it is possible to imagine schemes that would acquire many short exposure (underexposed) images, then to align and add them after acquisition. There would be a noise penalty, and this would extend the exposure time.

1.7.2 Zoom

Most conventional still and video digital cameras come with "zoom" optics and consumers have come to expect all digital cameras to have zoom lenses, sometimes with very large zoom ratios. The user can change the EFL of the taking lens continuously, which changes the size of the image on the sensor. MCMs could also, in principle, have zoom optics, but in most products today (mobile phones, tablets, and laptop computers) the space and cost constraints force fixed focal length optics and the optical zoom technology has not yet (2013) been widely adopted for MCMs.

Conventional optical zooms work by moving groups of optical elements along the optical axis. Usually, two motions are necessary, because one motion will typically change the focal length and it will also change the focus setting. A second moving group is needed to restore the focus. The optical design of a zoom system is often quite complex, requiring more components, and the relative aperture (fno) is slower than a nonzoom design might achieve. A zoom design may cause the exit pupil to shift axially, which should be considered when working with a "microlensed" digital sensor.

Miniature versions of conventional zoom lens designs would have more lens elements, at least two moving or tunable groups, and they would occupy much more volume within the product. Nevertheless, optical zooming is a desirable feature, and the work being done today may be found in future products. Meanwhile, an MCM has been introduced that has an unusually high-resolution sensor and a highly corrected nonzoom lens, maximizing the "digital zoom" capability in a mobile-phone camera.

1.7.3 Stereo 3D Photography

Still photography has been extended to stereo 3D for many years, sometimes with great success in the marketplace. In recent years, 3D motion pictures have become much more popular, so there is increased interest in stereo 3D. Some consumer cameras and video cameras produce 3D images, and stereo 3D is a natural application for MCMs.

Stereo 3D is the familiar idea of recording two independent images, with separated entrance pupils, and displaying those images separately to the viewer's two eyes. If done well, the viewer perceives the depth of a natural scene. We will briefly describe the conditions for stereo 3D.

The display of 3D images is an important component of the system. We will not discuss it in detail here, but we will assume that the images are seen at a comfortable focus distance, with matching magnification and little distortion.

The recording cameras should ideally be identical, aimed parallel to each other, and simply shifted by a distance that we will call the stereo baseline.

Three first-order parameters of the recording/display systems are important (Figure 1.26):

1. *Angular magnification*:

$$Ma = \frac{\tan\left(\text{viewed object angle}\right)}{\tan\left(\text{recorded CRA}\right)} = \frac{\text{Ubar}'}{\text{Ubar}}$$

2. *Stereo baseline magnification*:

$$Mp = \frac{\text{Viewer's pupil distance or eye separation}}{\text{Recorded stereo baseline}} = \frac{P'}{P}$$

3. *Apparent infinity distance*:

$$Z_{\text{inf}} = \frac{P'}{(2U')}$$

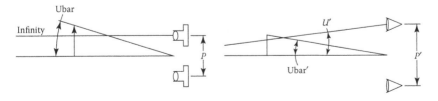

FIGURE 1.26 Schematic demonstration of the stereoscopic recording concept.

If we know these three parameters, we can understand the imaging of the 3D space.

Notice that the focus of the acquisition cameras and the display system does not change the 3D relationships; only the sharpness is changed. One fundamental difference of viewing a stereo 3D image is that while the 3D volume is reproduced, the *focus* of the images is not 3D; the original images are conventional photographs, with limited, fixed depth of field, and the display system gives us one fixed apparent best focus distance, regardless of the apparent depth of the object. This focus "disparity" does not usually disturb the illusion of reality for the viewer.

If the viewed volume is to be a 1:1 reproduction of the object volume (a condition called "orthostereo"), all three of the first-order parameters must be constrained: The two magnifications must be unity and the infinity convergence must be zero. If the angular magnification changes, the imaged volume will be stretched or compressed. If the stereo baseline magnification changes, the reproduced volume will be uniformly scaled up or down in x, y, and z. If the infinity convergence is not zero, x, y, and z are not reproduced accurately, but the scene is usually perceived as undistorted, because the human response seems to be based on angle differences, not the true angles; the brain decides that the most distant plane is infinity, and we interpret the rest of the scene based upon that assumption.

It is interesting to note that if only the stereo baseline magnification is varied from the orthostereo case, the result is a uniform magnification change. For example, if the camera baseline is half of the user's eye baseline, $Mp = 2$, and the reproduced volume appears 2× magnified to the user; a $2 \times 2 \times 2$ cube 10 cm from the camera will appear to be a $4 \times 4 \times 4$ cube 20 cm from the viewer. Likewise, large baseline photography can make a cityscape look like a miniature model.

When designing a stereo 3D system, some human factors should be considered to ensure comfortable viewing. First, although humans are tolerant of some error in horizontal convergence, angular disparity in the vertical direction should be avoided. Vertical disparity can obviously be caused by camera or display-aiming errors, but it can also be the result of a left–right mismatch of magnification, keystone distortion caused by nonparallel camera axes, unequal rotation of images about the optical axis, or rotation of the viewer's eye baseline from the display baseline. Some of these concerns become more serious in systems with very large display FOV.

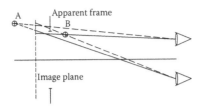

FIGURE 1.27 Geometry of the stereo frame.

As mentioned, focus errors can be less critical, but for comfortable viewing, we should try to keep the apparent display distance reasonable. If it is not adjustable, aiming for 1–2 m (1/2 to 1 diopter) would work well for most people.

We have not mentioned the "frame" created by the edges of the stereo photograph (Figure 1.27). The frame also appears at an apparent distance, which might be closer or farther than the subject of the photograph. It may seem more natural to the viewer if the frame is as close as the nearest part of the image, giving the feeling of looking through a window at the scene. If the image field is rectangular, the frame distance may be adjusted simply by cropping the left and right edges of the two channels. Remember that, in the strange world of stereo 3D, the binocular distance of the frame is independent of its focus distance, which is the same as the rest of the image.

It is easy to imagine the broad application of MCMs to stereo acquisition, because of their low cost and small size. The challenge is more to provide convenient display methods.

References

Bareau, J. and Clark, P.P. (2006), The optics of miniature digital camera modules, in G.G. Gregory, J.M. Howard, and R.J. Koshel (eds), *International Optical Design Conference, Proceedings of SPIE 6342*, 63421F, SPIE Press, Bellingham, WA.

Brady, D.J. and Hagen, N. (2009), Multiscale lens design, *Opt. Express* **17**(13), 10659–10674.

Fiete, R.D. (2012), *Formation of a Digital Image: The Imaging Chain Simplified*, SPIE Press, Bellingham, WA.

Gallegos, D., Long, K., Yu, H., Clark, P.P., Lin, Y., George, S., Nath, P., and Cunningham, B.T. (2013), Label-free biodetection using a smartphone, *Lab Chip* **13**, 2124–2132.

Goodman, J.W. (2005), *Introduction to Fourier Optics*, Roberts & Co., Greenwood Village, CO.

ISO 12233 (1998), Photography—Electronic still picture cameras—Resolution measurements, ISO.

ISO 14524 (2009), Photography—Electronic still-picture cameras—Methods for measuring opto-electronic conversion functions (OECFs), ISO.

ISO 9039 (2008), Optics and photonics—Quality evaluation of optical systems—Determination of distortion, ISO.

ISO 9358 (1994), Optics and optical instruments—Veiling glare of image forming systems—Definitions and methods of measurement, ISO.

Kingslake, R. (1983), *Optical System Design*, Academic Press, New York.

Kingslake, R. and Johnson, B. (2009), *Lens Design Fundamentals*, Academic Press, San Diego, CA.

Malacara, D. (ed.) (2007), *Optical Shop Testing*, Wiley-Interscience, Hoboken, NJ.

Ng, R. (2006), Digital light field photography, PhD dissertation, Stanford University, https://www.lytro.com/renng-thesis.pdf.

Plummer, W.T., Baker, J.G., and Van Tassell, J. (1999), Photographic optical systems using nonrotational aspheric surfaces, *Appl. Optics* **38**(16), 3572–3592.

Smith, W.J. (2007), *Modern Optical Engineering*, 4th edn, McGraw-Hill Professional, New York.

Zhou, G., Yu, H., and Chau, F.S. (2012), Microelectromechanically-driven miniature adaptive Alvarez lens, *Opt. Exp.* **21**(1), 1226–1233.

CHAPTER **2**

Modern Image Sensors

Junichi Nakamura

Contents

2.1 Introduction

A solid-state image sensor is a semiconductor device that converts the optical image formed by an imaging lens into electronic signals, as shown in Figure 2.1

From its commercialization in the early 1980s, the charge-coupled device (CCD) image sensor was the technology of choice for a long time for almost all imaging applications. However, recently, the complementary metal-oxide-semiconductor (CMOS) image sensor has taken that position.

In 1967, G. Weckler of Fairchild Research and Development Laboratory proposed a charge integration mode, where a signal charge is accumulated on an electrically floating photodiode over a period of one frame time (Weckler 1967). This concept triggered the development of metal-oxide-semiconductor (MOS)-type image sensors.

In 1969, W. S. Boyle and G. E. Smith of AT&T Bell Labs invented the CCD (Boyle and Smith 1970). Several applications for the CCD, such as analog memory and analog domain signal processing, were devised, but soon its use as an image sensor became the most promising application.

In the 1970s and the 1980s, the MOS image sensor competed with the CCD image sensor. The CCD image sensor required a sophisticated process technology to realize very high charge transfer efficiency, which was a big challenge. On the other hand, it was considered that the MOS image sensor was easier to build and would be a good match with the ever-growing very-large-scale integration (VLSI) process technology. In 1981, the first consumer color video camera was released with a MOS image sensor, prior to the release of a color CCD camera.

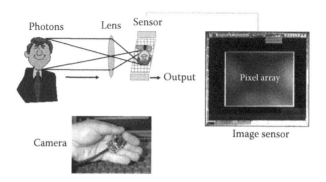

FIGURE 2.1 Schematic demonstration of the image sensor's operation in a mobile camera.

However, as the CCD technology evolved and its superior noise performance became evident over its MOS image sensor counterpart, the MOS image sensor disappeared in the early 1990s.

The primary factors that made the CCD image sensor become the technology of choice in imaging applications include (Teranishi 2000): the two-phase CCD (Krambeck et al. 1971), the buried channel CCD (BCCD) (Walden et al. 1971), the floating diffusion amplifier (FDA) (Kosonocky and Carnes 1971), correlated double sampling (CDS) (White et al. 1974), the pinned photodiode (PPD) (Teranishi et al. 1982), the vertical overflow drain (VOD) (Ishihara et al. 1982), and the on-chip color filter array (CFA)/microlens array (Ishihara and Tanigaki 1983).

In 1993, E. R. Fossum of NASA's Jet Propulsion Laboratory proposed a "camera-on-a-chip" concept using a CMOS active pixel sensor (CMOS APS) (Fossum 1993). The U.S. semiconductor industry quickly responded to this proposal and led early developments, while CCD companies around the world were skeptical about the CMOS APS concept. With continued improvement, CMOS image sensors found ever-wider use and it was projected that the CMOS image sensor would continue to dominate the imaging market.

In this chapter, the basics of the modern image sensor will be reviewed. The two most popular types of image sensors, the interline transfer CCD (IT-CCD) image sensor (Amelio 1973) and the four-transistor (4T) PPD CMOS image sensor, will be used as examples to explain state-of-the-art sensor operation and performance.

2.2 Modern Image Sensors

2.2.1 IT-CCD Image Sensors

There are three representative CCD image sensor architectures: (1) frame transfer (FT) CCD, (2) IT-CCD, and (3) frame interline transfer (FIT) CCD. The IT-CCD is used here to explain CCD technology, since it is the most popular scheme for consumer imaging applications. For the other two architectures, see Theuwissen (1995) and Yamada (2006).

The simplified sensor architecture of the IT-CCD image sensor is shown in Figure 2.2.

V-CCD is the vertical CCD register where the signal charge is transferred by the four-phase vertical charge transfer pulses,

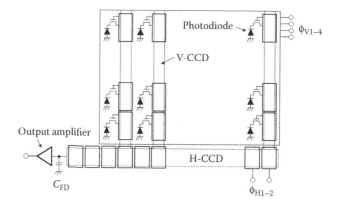

FIGURE 2.2 Interline transfer CCD.

ϕ_{V1-4}. H-CCD is the horizontal CCD register where the charge is transferred by the two-phase horizontal charge transfer pulses, ϕ_{H1-2}. C_{FD} is the charge sense node (floating diffusion, FD) capacitance.

In the IT-CCD image sensor, the signal charge that is generated and accumulated during a frame time at each photodiode is transferred to the V-CCD register simultaneously (i.e., global shutter) and is shifted toward the H-CCD register. Once the signal charges of a row of pixels reach the H-CCD, they are transferred along the H-CCD register and converted to a voltage at the output amplifier.

Figure 2.3 shows a basic charge transfer mechanism using the case of the four-phase CCD that is used for the V-CCD in the

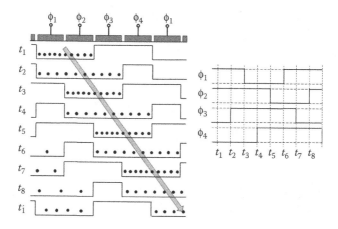

FIGURE 2.3 Charge transfer mechanism.

IT-CCD image sensor as an example. A high-level pulse creates a potential well in which the signal charge is stored. The signal charge is transferred from the left to the right by applying the four-phase pulses appropriately.

The charge detection scheme at the output stage of the IT-CCD image sensor is shown in Figure 2.4. The output amplifier usually consists of a two-stage or three-stage source follower amplifier. The charge sense node is denoted by FD, of which the charge-to-voltage conversion capacitance is C_{FD}.

At $t = t_1$, the signal charge that has reached the final transfer stage in the H-CCD register (here, a buried channel, two-phase CCD register is shown) and the FD is reset by the reset pulse ϕ_{RST}. When the reset pulse goes low, the output level is decreased by the capacitive feedthrough and the kTC noise appears on the FD (which is shown as a gray portion). Then, at $t = t_3$, when the pulse ϕ_2 becomes low, the signal charge stored at the final H-CCD stage is transferred to the FD. The output swing is given by

$$V_{sig} = A_V \cdot \frac{Q_{sig}}{C_{FD}} \tag{2.1}$$

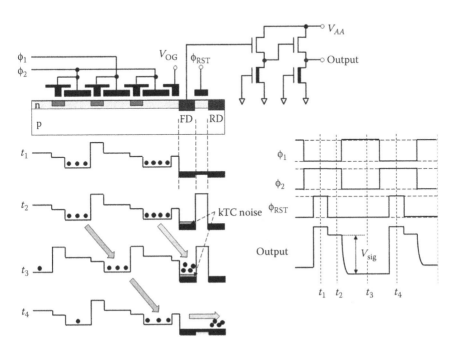

FIGURE 2.4 Charge detection scheme.

where Q_{sig}, A_V, and C_{FD} are the signal charge, the voltage gain of the output amplifier, and the FD capacitance, respectively.

This signal charge is swept to the reset drain (RD) when the reset pulse is applied again at $t = t_4$.

kTC noise, which originates from the thermal noise of a MOS transistor's on-resistance and appears when a MOS switch is turned off at a capacitor that connects to the MOS switch, is commonly suppressed by CDS (White et al. 1974), where the output signal at $t = t_2$ is subtracted from the output signal at $t = t_3$.

Figure 2.5 shows the typical pixel structure of the IT-CCD image sensor. The photodiode is a PPD where an n layer is sandwiched by a grounded surface p^+ layer and p-well. With this structure, it is possible to fully deplete the n-region and realize a complete charge transfer from the photodiode to the V-CDD register. Also, a very low dark current performance is obtained because the surface p^+ layer fills the interface states that would generate dark currents.

An antireflection layer is commonly used to enhance quantum efficiency. The n-type substrate is positively biased (typically +8 V or so), which helps suppress pixel-to-pixel cross talk, since electrons generated by long wavelength photons that would diffuse into neighboring pixels are absorbed in the n-substrate. Also, it serves as a VOD, as shown in Figure 2.6, where the excess charge generated by very strong incident light is drained to the n-substrate, which, in turn, prevents blooming. When a high voltage is applied to the n-substrate, the signal charge stored in the photodiode is drained to the n-substrate. This operation is used for the electronic global reset shuttering.

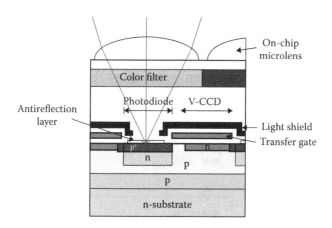

FIGURE 2.5 Pixel structure of the IT-CCD.

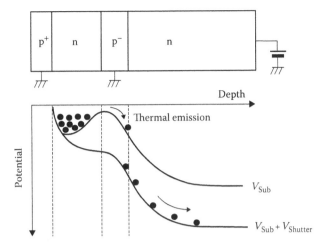

FIGURE 2.6 Principle of the vertical overflow drain and the electronic global shutter.

2.2.2 CMOS Image Sensors

Figure 2.7 shows a digital-output CMOS image sensor, where analog-to-digital conversion (ADC) takes place on-chip. It is possible to put more digital signal processing circuitry on a CMOS image sensor because of the CMOS fabrication process, making the CMOS image sensor a "camera on a chip" (Fossum 1997).

Figure 2.8 is a diagram of the pixel structure and the column sample-and-hold circuit of the 4T CMOS image sensor. Note that there are other pixel structures, such as the three-transistor photodiode pixel and the 4T photo-gate pixel (Fossum 1997; Takayanagi 2006). Here, we will focus on the 4T PPD pixel and its derivative pixels.

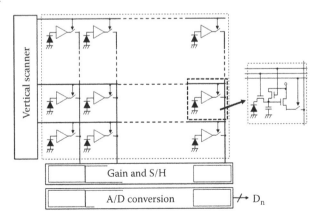

FIGURE 2.7 Digital-output CMOS image sensor.

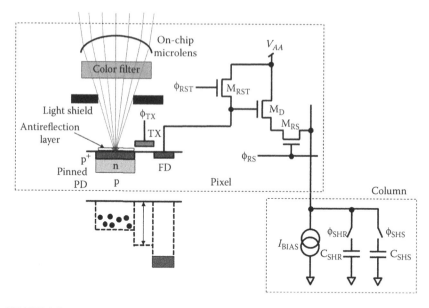

FIGURE 2.8 Pixel configuration and a column sample-and-hold circuit in the CMOS image sensor.

The 4T pixel consists of a PPD; the transfer gate, TX; the driver transistor, M_D; the row-select transistor, M_{RS}; and the reset transistor, M_{RST}. A source follower amplifier is formed by the driver transistor, M_D, and a load transistor, which is located outside the pixel array and is represented as a current sink, I_{BIAS}, in this figure, through the row-select transistor, M_{RS}.

The signal charge accumulated on the photodiode is transferred to the charge sense node, which is the floating diffusion (FD), by pulsing the transfer gate, TX, and is converted to a voltage. The signal voltage is then fed to a column signal processing circuit, which will be described later, and the analog output signal is converted to a digital code.

The pixel of the CMOS image sensor is an "active pixel" in that the charge-to-voltage conversion is performed inside the pixel, which can enhance the signal-to-noise ratio (SNR). However, a variation in the threshold voltage of the driver transistor, M_D, causes pixelwise fixed-pattern noise (FPN). An FPN suppression circuit is implemented on-chip in the modern CMOS image sensor.

The signal voltage swing on the column line (or the vertical signal line) is also given by Equation 2.1, where A_V is the voltage gain of the pixel source follower amplifier. As can be understood from Figure 2.7, the charge-to-voltage conversion and the signal voltage readout are done on a row-by-row basis. This signal voltage is amplified by a column programmable gain amplifier, processed for FPN suppression, and is converted to a digital code by on-chip ADC. The ADC can take place either on a column level (column-parallel architecture, where multiple ADCs operate in parallel during a row time) or on a chip level (a serial architecture, where a single A/D converter operates at a pixel rate).

An example of the FPN suppression circuit and its timing diagram is shown in Figure 2.9a and 2.9b, respectively. After an integration time, the floating diffusion, FD, is reset by pulsing the reset transistor, M_{RST}, on. The kTC noise appears at the FD node and a corresponding pixel output "reset" voltage, V_{col_RST}, is sampled on a sample-and-hold capacitor, C_{SHR}, by a pulse, ϕ_{SHR}. Then, the signal charge generated and stored in the PPD is transferred to the FD by the transfer pulse, ϕ_{TX}. A corresponding pixel output "signal" voltage, V_{col_SIG}, is sampled on a sample-and-hold capacitor, C_{SHS}, by a pulse, ϕ_{SHS}.

The pixelwise FPN can be suppressed by subtracting V_{col_RST} from V_{col_SIG}, as the identical threshold voltage variation is contained in these two samples, as shown in Figure 2.9c.

In the following sections, we will describe how sensor performance items are affected by the sensor architecture.

2.2.3 Sensor Performance Determined by Sensor Architecture and/or Operating Mechanism

2.2.3.1 Global Shutter Versus Rolling Shutter

As shown in Figure 2.10, the IT-CCD realizes the global shutter operation, where the photodiode reset and the signal charge readout (both are done by the charge transfer from PPD to V-CCD)

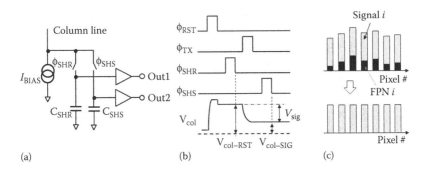

FIGURE 2.9 FPN suppression: (a) FPN suppression circuit, (b) timing diagram, and (c) output.

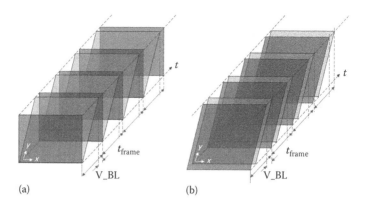

FIGURE 2.10 (a) Global shutter and (b) rolling shutter.

occur at the same time for all pixels. However, the signal volt-ages from the output amplifier appear row by row (the charge transfer from V-CCD to H-CCD is done on a row-by-row basis) through a signal charge transfer in the H-CCD register. This mechanism is shown in Figure 2.10a, where the dark gray planes represent the photodiode reset and the signal charge readout to the V-CCD (the signal charge is accumulated for a period that corresponds to the time between two gray planes) and the pale gray planes represent the signal voltage readout from the output amplifier.

On the other hand, the shutter operation of the CMOS image sensor is called the "(electronic) rolling shutter," because the charge reset, charge integration (accumulation), and signal readout are per-formed on a row-by-row basis, as shown in Figure 2.10b.

Image distortion is generated for a moving object when a still image is captured and when a recorded frame rate and the frame rate of a video display are not identical, because of the row-by-row imaging operation.

Figure 2.11 shows reproduced images of (a) a global shutter and (b) a rolling shutter, at 45 frames per second (fps). With the global shutter, there is no distortion of the blades of the fan, but this is quite evident with the rolling shutter.

Other disadvantages of the rolling shutter are (1) a strobe can only be used when all photodiodes are in the charge accumula-tion period (no reset/readout scans are being applied), and (2) an unsynchronized flash/strobe might cause a bright band in the resultant image.

(a)   (b)

FIGURE 2.11 Two images showing the difference between the global shutter and the rolling shutter when photographing a moving object. 45 fps, t_{int} = 4 ms: (a) global shutter and (b) rolling shutter.

2.2.3.2 Frame Rate

The speed limitation of the CCD image sensor obviously comes from its charge transfer readout mechanism where the charge packet stored in the potential well must be transferred to the next potential well. The reported highest H-CCD clock frequency of consumer-use IT-CCD image sensors is around 50 MHz. If we take the digital HDTV 1080p standard (Recommendation ITU-R 2002) as an example where the number of effective columns N_H is 1920, the number of effective rows N_V is 1080, the frame rate is 59.94 fps, and the line frequency is 67.5 kHz (one row time of 14.8 μs), a clock frequency greater than 150 MHz is required for the H-CCD. Also, there is a speed limitation for the CDS to work properly. Multiple H-CCD channels are thus required to meet the HDTV specifications.

As the pixel size is reduced to achieve higher spatial resolution, the area for the V-CCD in a pixel is also reduced, which reduces the charge-handling capacity of the V-CCD. A multiple interlace readout scheme has been introduced to maximize the full-well capacity of the photodiode, while utilizing multiple V-CCD stages for one photodiode (e.g., 12:1 interlaced fields form one frame image). This scheme further slows down the frame rate.

Next, we will examine the frame rate in the CMOS image sensor. Let us take the same example: a 1080p 60 fps mode. Achieving the 150 MHz scan rate is also a challenge for an analog-output CMOS image sensor. The solution is the same as its CCD counterpart, which is to implement multiple readout channels to reduce the analog scan rate for each channel. However, this approach increases the power consumed by output buffers and complicates analog front-end circuitry.

A digital-output CMOS image sensor can solve this issue. In the digital-output CMOS image sensor, on-chip ADCs are implemented and a column-parallel architecture (signal processing for pixel output signals is done in parallel by column-parallel circuits during a row time), as shown in Figure 2.7, in particular, can offer a high frame rate and low power solution.

Once the analog signal is converted to digital codes, handling the 150 MHz data stream is not a big challenge and, additionally, a lower digital domain voltage supply can be used. Several column-parallel ADC schemes have been reported and A/D conversion times of 2–3 μs are available for a 12–14 bit resolution. A signal processing time of a few microseconds is feasible for amplifying the pixel analog signal and FPN suppression.

Also, contrary to the charge transfer scheme in the CCD image sensor, the CMOS image sensor uses metal wiring to read out the pixel output signal, which permits faster and more flexible readout schemes, such as a windowing readout or a skip readout.

2.2.3.3 Power Consumption

In the IT-CCD image sensor, the power to drive the CCD registers is given by

$$P_{CCD} = \frac{1}{2} \cdot \sum C_i \cdot V_i^2 \cdot f_i \qquad (2.2)$$

where C_i, V_i, and f_i denote the total load capacitance for the clock i (e.g., $\phi_{V1} \sim \phi_{V4}$, $\phi_{H1} \sim \phi_{H2}$, etc.), the voltage swing of the clock i, and the frequency of the clock i, respectively.

Also, power is consumed by the output amplifier, which is given by

$$P_{amp} = I_{BIAS} \cdot V_{AA} \qquad (2.3)$$

where V_{AA} is the supply voltage for the output amplifier (which is typically 15 V) and I_{BIAS} is the bias current of the output amplifier, which is determined from a settling time requirement for the output waveform.

From Equation 2.2, it can be understood that P_{CCD} increases with the clock frequency, meaning more power is needed for a larger format, or a higher-resolution image sensor. The same argument can be applied to P_{amp} because higher I_{BIAS} is required.

In the CMOS image sensor, a much lower supply voltage is used, such as 3 V for the analog supply and 1.8 V for the digital supply, or even lower with an advanced process technology. Also, a low-voltage differential signaling (LVDS) scheme for the output stage, instead of the CMOS-level parallel output scheme, is commonly used, which helps reduce the power consumption.

In addition, only one row of the pixel array is activated in a frame due to its row-by-row operation sequence. Therefore, the power to drive the pixel array is much lower than that for the CCD image sensor.

Total power consumption can vary, depending on the sensor architecture, ranging from a simple analog-output architecture to a system-on-chip (SOC)-type architecture. In any case, power consumption (and how much on-chip circuitry should be implemented) should be assessed from a system point of view.

2.2.3.4 Noise Performance

Let us consider the noise performance of modern image sensors by referring to Figure 2.12, where simplified signal chain diagrams for the IT-CCD image sensor and the CMOS image sensor are

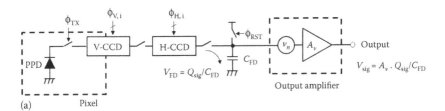

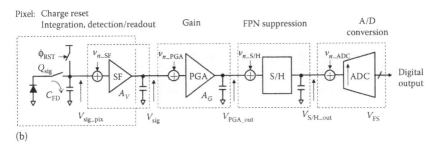

FIGURE 2.12 Signal chain diagrams: (a) signal chain of the CCD image sensor and (b) signal chain of the digital-output CMOS image sensor. Note that this scheme is just one example and modifications to optimize the signal chain are possible.

shown. Here, we focus on the circuit noise, which is referred to as the noise floor or read noise.

In the IT-CCD image sensor (Figure 2.12a), only one random (temporal) noise source exists at the output amplifier, which includes thermal noise, $1/f$ noise of the output amplifier, and the kTC noise associated with a reset of the FD node. The complete charge transfer does not generate any noise, which makes the CCD image sensor a remarkably low-noise electronic component.

Assuming that the kTC noise and the $1/f$ noise can be suppressed by the CDS, the output noise (i.e., noise measured at the output), v_{n_out}, is represented by

$$v_{n\_out}^2 = 2 \cdot A_V^2 \cdot v_n^2 \qquad (2.4)$$

where A_V and v_n are the voltage gain of the output amplifier and the input-referred thermal noise component of the output amplifier, respectively, and the factor of 2 comes from the CDS operation. The number of input-referred noise charges (i.e., the noise charge at the FD node, calculated using the parameters shown in Figure 2.12) is given by

$$n_{CCD} = \sqrt{2} \cdot \frac{C_{FD}}{q} \cdot v_n \qquad (2.5)$$

where q is the unit charge (1.6×10^{-19} [C]).

As previously mentioned, for the CDS circuit to work properly, it must remain below a certain maximum frequency (at frequencies above that maximum, its performance degrades).

State-of-the art IT-CCD image sensors for consumer use offer 2–5 e$^-$ noise, that is, the root mean square (RMS) input-referred noise corresponds to 2–5 electrons.

It is possible to reduce the output noise by lowering the output rate and applying low-pass filtering to the output signal. This technique is used in scientific applications and the resulting output noise is proportional to the square root of the frequency bandwidth.

Several noise sources exist in the CMOS image sensor as shown in Figure 2.12b. The first noise source is the pixel source follower transistor. Noise from this transistor includes thermal noise, $1/f$ noise, and random telegraph signal (RTS) noise. The RTS noise appears as a flickering noise in small MOS transistors and is caused by the temporal capture and emission of channel electrons between the transistor channel and the gate oxide, thus having two values over time. With the 4T pixel configuration, the $1/f$ noise component can be suppressed together with the FPN (caused by a threshold voltage variation of the source follower transistor) by the CDS operation. However, the CDS operation doubles the thermal noise component, as expressed by Equation 2.4, and causes the RTS noise to have three values.

The input-referred noise charge (i.e., the calculated noise charge at the pixel FD node) of a digital-output CMOS image sensor is given by

$$n_{\text{CMOS}} = \sqrt{2} \cdot \frac{C_{\text{FD}}}{q} \cdot \sqrt{v_{n\_\text{SF}}^2 + \frac{v_{n\_\text{PGA}}^2}{A_V^2} + \frac{v_{n\_\text{S/H}}^2 + v_{n\_\text{ADC}}^2}{A_V^2 \cdot A_G^2}} \quad (2.6)$$

where A_V, v_{n_SF}, A_G, v_{n_PGA}, $v_{n\_\text{S/H}}$, and v_{n_ADC} are the voltage gain of the pixel amplifier (source follower), noise of the pixel amplifier, the voltage gain of the PGA, noise of the S/H circuit, and noise of the ADC, respectively. It can be seen that the noise floor is reduced with the PGA gain, as shown in Figure 2.13.

State-of-the art CMOS image sensors can offer <2 e⁻ noise by applying high column gain A_G, while the handling signal range is reduced by a factor of A_G.

Noise performance is not dependent on the output pixel rate in the column-parallel architecture, since the A/D conversion is

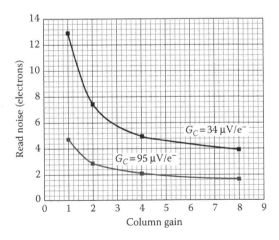

FIGURE 2.13 An example of noise performance of the CMOS image sensor.

complete at the column-parallel circuitry before reading out the resulting digital codes.

Until recently, FPN was a serious problem in CMOS image sensors (note that a CCD image sensor has no FPN). Due to this FPN, high gain settings could not be used in CMOS cameras, resulting in low-sensitivity cameras. However, thanks to recent progress in FPN suppression, FPN has been brought to a level low enough for practical use (see Section 2.5).

From the previous discussions, the operating speed, power consumption, and noise performance are highly linked. Therefore, design optimization is necessary for each particular application. However, in general, CMOS image sensors have inherent advantages over CCD image sensors for high-resolution image sensors by providing high frame rates with less power consumption (Takayanagi and Nakamura 2013).

2.2.4 Pixel Size Reduction

2.2.4.1 Trend of Pixel Size Reduction

Digital cameras, in general, have evolved to offer higher resolution (higher pixel counts) and a smaller and lighter camera body. The optical format refers to the size of the imaging array area of the image sensor and is more or less fixed for particular applications, such as 1/1.7″ (7.5 × 5.6 mm), 1/2.3″ (6.3 × 4.7 mm) or smaller for compact point-and-shoot digital still cameras (DSCs) and smartphone cameras, 1/3″ (4.8 × 3.6 mm) or smaller for consumer video cameras, and full size (36 × 24 mm), APS-C (23 × 15 mm), 4/3″ (17.3 × 13 mm), 1″ (13.2 × 8.8 mm), or 2/3″ (8.8 × 6.6 mm) for digital single-lens reflex (DSLR) cameras and/or "mirrorless"/interchangeable lens cameras. In any case, the pixel size needs to be reduced to increase the spatial resolution. The relationship between the pixel size and the number of pixels with the optical format as a parameter is shown in Figure 2.14. Note that recent image sensors use a square pixel. Thus, the pixel size is the pixel pitch squared.

Also, the use of "type" is recommended to express the optical format, since the use of "inch" is confusing as 18 or 16 mm (instead of 25.4 mm) is used conventionally to express the diagonal length of the image area from the era of vacuum tubes.

Figure 2.15 shows the trends of pixel size reduction for both the CCD and the CMOS image sensors. In both cases, the use of progressively more advanced/finer process technologies brought about smaller pixels; however, for CMOS image sensors, the

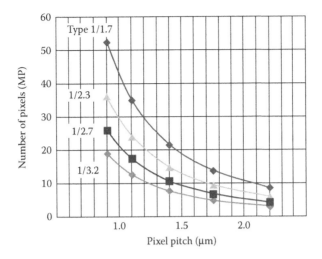

FIGURE 2.14 The relationship between pixel pitches and the number of pixels for given optical formats (for compact DSC/mobile phone camera applications).

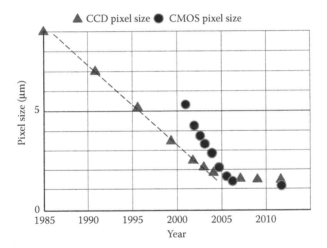

FIGURE 2.15 Trend of pixel size reduction.

use of a "shared pixel" scheme, which cannot be used in CCD image sensors, contributed to a more rapid reduction in pixel size. However, since around 2005, progress in pixel size reduction for both CCD and CMOS has slowed down (1.4, 1.1, and 0.9 μm).

2.2.4.2 Shared Pixel Scheme in the CMOS Image Sensor

So, what is the shared pixel scheme? As shown in Figure 2.8, the pixel of the CMOS image sensor has four transistors, which might make one think that it would be difficult to shrink the pixel more rapidly than the progress of the fabrication technology. However, a clever idea was suggested: that the three transistors M_{RST}, M_D, and M_{RS} be used for several photodiodes, as shown in Figure 2.16a. This scheme is called a "shared pixel" scheme and it can be realized because the photodiode and the three transistors are separated by the transfer gate, TX. Although Figure 2.16 shows an example of a two-row shared scheme, other shared schemes, such as a four-shared scheme, where four photodiodes (2 (row) × 2 (column) or 4 (row) × 1 (column)) or more (e.g., 8 (row) × 1 (column)) are shared with one readout/reset circuitry, are available. The effective number of transistors per pixel becomes 2.5 for the two-row shared pixel (5 transistors/2) and 1.75 for the four-shared pixel (7 transistors/4), and so on.

Figure 2.16b and 2.16c show the different placement examples of the two-row shared scheme. The sense node FD capacitance, which determines the conversion gain, is small in the configuration shown in Figure 2.16b, while a performance mismatch between the two photodiodes could appear due to their unsymmetrical layout. The configuration shown in Figure 2.16c, where an identical layout is used for two photodiodes and FDs (which are connected together by wiring), addresses this point at the expense of increased FD capacitance (i.e., reduced conversion gain). The shared pixel scheme has contributed to a pixel size reduction below ~3 μm.

2.2.4.3 FD Drive: Removal of Row-Select Transistor in CMOS Pixel

The removal of the row-select transistor, M_{RS}, is something else that can be done to reduce the size of the pixel, or to make it

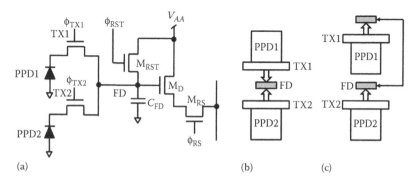

FIGURE 2.16 Shared pixel configuration: (a) shared pixel configuration, (b) layout 1, and (c) layout 2.

possible to enlarge the photodiode area, in CMOS image sensors. If the charge sense node (FD) potential of the pixels on nonselected rows is set below the threshold voltage of the source follower transistor, M_D, only the pixel values on the selected row appear on the column lines, even though the row-select transistor is not implemented in a pixel.

The FD potential is controlled through the reset transistor, M_{RST}, and there are several configurations to apply the target potentials to the selected row and the nonselected rows.

The effective number of transistors per pixel becomes 2 for the two-row shared pixel (4 transistors/2) and 1.5 for the four-shared pixel (6 transistors/4), and so on.

2.2.4.4 Backside Illumination

One of the shortcomings of the conventional pixel used in the CMOS image sensor compared to that of the pixel used in a CCD image sensor is that the distance between the bottom of the microlens and the surface of the photodiode (stack height) is larger. This is because multiple metal layers are used in the CMOS pixel and this causes stronger dependency on the lens F-number and the angular response compared to the CCD image sensor (see Section 2.5.2).

To solve this issue, sensors that are illuminated from the "back side" (backside illumination, BSI) have been developed and are in practical use (Wakabayashi et al. 2010). Figure 2.17 shows a schematic cross-sectional diagram of the BSI CMOS image sensor. The conventional scheme is now called frontside illumination (FSI). The BSI image sensor is fabricated as follows: after the metal processing of an FSI CMOS image sensor, a support wafer ("mechanical substrate" in Figure 2.17) is attached; the

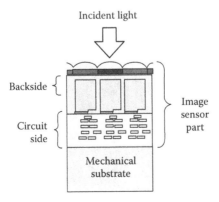

FIGURE 2.17 BSI (backside illumination) scheme.

sensor wafer is ground to a thickness of a few micrometers, and a CFA and a microlens array are formed on what was originally the backside. This results in an increase in the responsivity and an improvement in the angular response, and it allows a large amount of freedom for metal routing.

2.2.4.5 Light Guide

The extra manufacturing steps required for BSI sensors make them more expensive than FSI image sensors. Other improvements in the manufacturing process to suppress cross talk between pixels and dark currents are needed with additional cost.

Consequently, FSI image sensors with a light guide or light pipe, as shown in Figure 2.17, have been developed (Agranov et al. 2011). Light guides are formed following the topmost metal process by etching the insulator layer on top of the photodiode and injecting a polymer with a larger refractive index than the insulator layer. The on-chip microlenses are designed to concentrate light from the imaging lens on the top of the light guide. If there were no light guides, light with a large angle of incidence would enter the neighboring pixels causing pixel-to-pixel cross talk, as illustrated in the left plot in Figure 2.18. As shown in the scanning electron microscope (SEM) photograph in Figure 2.18, the effective stack height is reduced to h_2 from h_1, which is the stack height without the light guide.

2.2.4.6 Diffraction Limit

The imaging lens of a camera is not perfect in that a point light source cannot be focused at a point on the focal plane due to the wave characteristic of light. The diameter of the Airy disk, D, which induces the lens diffraction limit, is given by

$$D = 2.44F\lambda \tag{2.7}$$

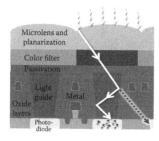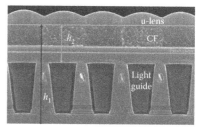

FIGURE 2.18 Light guide.

where F and λ are the lens F-number and the wavelength of the light, respectively. According to the Rayleigh criterion, two separate point images (of diameter D) can be resolved if the distance between the peaks of these two point images (L in Figure 2.19a) is greater than $D/2$. If this criterion is used to consider the minimum pixel size, then the minimum pixel size is given by

$$p = 0.61F\lambda \tag{2.8}$$

assuming the Bayer color filter pattern (Figure 2.19b; Bayer 1976). Figure 2.19c shows the relationship between p and the lens F-number. With $F2.8$, it is anticipated that the minimum pixel size is ~ 0.9 µm, although a more rigorous analysis is needed to precisely identify the minimum pixel size.

In order to go beyond diffraction-limited pixels, the concept of a quanta image sensor (QIS) has been proposed (Fossum 2012). The QIS consists of an array of billions of nanoscale, binary pixels, each having a binary output representing the absence ("0") or presence ("1") of a photoelectron. An image is obtained by

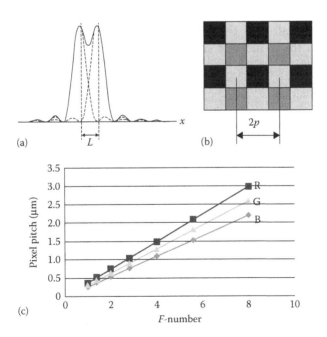

FIGURE 2.19 Diffraction-limited pixel pitch: (a) two point images, (b) Bayer CFA pattern, and (c) minimum pixel pitch vs. F-number.

manipulating massive data from the QIS, which are read into and stored in an external digital memory.

2.2.5 Pixel and Array Performance

2.2.5.1 Pixel Performance

a. Photoconversion characteristics

An example of photoconversion characteristics, the relationship between exposure (faceplate illuminance × charge integration time) and the number of generated electrons, is shown in Figure 2.20. The number of signal electrons and input-referred photoresponse nonuniformity (PRNU) is in proportion to exposure. The photon shot noise comes from the particle nature of the photon and is proportional to the square root of the signal electrons (and thus exposure). The noise floor, also called "read noise," is noise generated by readout circuitry and is not dependent on the exposure.

The number of signal electrons is represented by

$$N_{\mathrm{sig}} = R \cdot E \tag{2.9}$$

where R is the responsivity in electrons per lux-second (electrons/lx-s) and E is the exposure in lux-second. Responsivity and thus the number of signal electrons are determined only by the

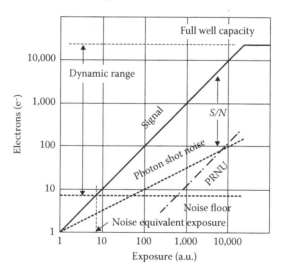

FIGURE 2.20 Photoconversion characteristics.

quantum efficiency of the pixel with a given light source and an infrared (IR)-cut filter. The signal voltage at the charge sense node (*FD*) is given by

$$V_{\text{sig}} = G_C \cdot N_{\text{sig}} \qquad (2.10)$$

where G_C is the conversion gain in electrons per volt (electrons/V) and is represented as

$$G_C = \frac{q}{C_{\text{FD}}} \qquad (2.11)$$

The full-well capacity is determined by the photodiode structure (size and impurity profile) and the voltage applied to the transfer gate (the transfer gate from PPD to the V-CCD in the IT-CCD image sensor and the transfer gate from PPD to FD in the CMOS image sensor).

In the digital-output CMOS image sensor, the maximum handling charge is usually limited by the input voltage of the on-chip ADC to the linear characteristic output (the photoconversion characteristic near saturation is not linear). In this case, the maximum number of signal electrons is smaller than the full-well capacity of the photodiode.

Sometimes, the term *sensitivity* is used instead of "responsivity" to indicate the quantity of electrons generated per unit of exposure. Another definition of sensitivity is how dark a scene can be taken by the image sensor. In this definition, both the responsivity and the noise performance must be considered. For example, noise equivalent exposure can be defined to express the "sensitivity," as shown in Figure 2.20, at which the number of signal electrons is equal to the number of noise electrons (SNR is 1). The dynamic range is defined as the ratio of the maximum handling charge to the noise floor.

The SNR is the ratio of the number of signal electrons to noise at a particular exposure level. For exposure levels where photon shot noise is dominant, the SNR is given by

$$\text{SNR} = \frac{N_{\text{sig}}}{\text{shot noise}} = \frac{N_{\text{sig}}}{\sqrt{N_{\text{sig}}}} = \sqrt{N_{\text{sig}}} \qquad (2.12)$$

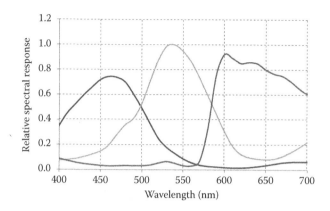

FIGURE 2.21 Example of spectral response.

Equation 1.12 implies that the number of signal electrons can be estimated by measuring the SNR.*

b. Spectral response

Most image sensors require a color filter to separate the colors and the most widely used color filter is the "Bayer" pattern (Bayer 1976). An example of the spectral response of a color image sensor is shown in Figure 2.21. The spectral response determines fidelity in color reproduction.

c. Performance at low illumination levels: dark current and white blemish

A dark current is an undesirable current that is integrated as a dark charge at a storage node inside a pixel. The amount of dark charge is proportional to the integration time, t_{int}, and is also a function of temperature T, as shown in Equation 1.13:

$$N_{dark}(t_{int}, T) \propto t_{int} \cdot \exp\left(-\frac{E_g}{\alpha kT}\right) \qquad (2.13)$$

where E_g is the bandgap energy of silicon, k is Boltzmann's constant, and α $(1 < \alpha < 2)$ is a coefficient that is related to the origin of the dark current.

---

* The definitions of camera sensitivity and camera dynamic range are different in the video camera engineering community; camera sensitivity is defined by the lens F-number at a predetermined level of subject illuminance (2000 lx is most often used), at which the camera output reaches the standard output level. The definition of camera dynamic range is the ratio of exposure values, one corresponding to the camera saturation level and the other corresponding to the camera white level.

The dark charge reduces an image sensor's dynamic range. It also changes the output level that corresponds to "dark" (no illumination). Therefore, optical black (OB) pixels, which are shielded from incident light by a metal layer and/or black material, are included in the pixel array and their output levels are clamped to provide a reference value for a reproduced image.

The PPD structure eliminates the surface dark current component that would be generated at the Si surface and it is the largest component of a dark current, unless a surface p^+ layer is introduced (see Figures 2.5 and 2.8).

Also, applying a negative voltage to the transfer gate during the integration time is effective to suppress a dark current that would be generated under the transfer gate, for both IT-CCD and CMOS image sensors. Likewise, in the IT-CCD image sensor, a negative voltage (e.g., -8 V for the low level of the vertical charge transfer pulses ϕ_V's) is applied for pulses to drive the V-CCD momentarily to reduce surface-oriented dark currents (Yamada 2006). These operations utilize holes to empty the surface generation centers (Theuwissen 2006).

White pixels or white blemishes appear as white spots in a reproduced image and detract from the quality of the image sensor. The causes of white blemishes include contamination by heavy metals and crystal defects induced by stress during fabrication. Although the state-of-the-art image signal processor (ISP) corrects a certain amount of white blemishes, fabrication technologies need to be improved to suppress these undesirable components.

d. Image lag

Image lag is a phenomenon in which a residual image remains in succeeding frames after the light intensity suddenly changes, either light to dark or dark to light. Image lag can appear even in PPDs if there is a bump or a dip in the potential along the charge transfer channel from the PPD to the V-CCD in the IT-CCD image sensor and to the FD in the CMOS image sensor, resulting in an incomplete charge transfer. Image lag also degrades linearity in the photoconversion characteristics at low output levels. Generally speaking, modern image sensors have solved this issue.

e. Cross talk

If there is signal cross talk between pixels, spatial resolution and color reproduction worsen. There are two types of cross talk: one is optical cross talk, which occurs when the incident light itself enters a neighboring pixel, and the other is electrical cross talk, which occurs when a signal charge generated deep in the Si bulk diffuses and enters neighboring pixels. As for the former, reducing the distance from the bottom of the microlens to the surface of the photodiode (stack height) and/or introducing light guides are effective. For the latter, if the signal charge is electron, using an n-type substrate can reduce the effective photoconversion depth while maintaining sufficient responsivity for the red region of the spectrum.

f. Performance under high illumination levels

Several artifacts that detract from image quality can appear under high illumination levels, such as smear and blooming. Smear is a CCD image sensor-specific phenomenon and is a bright vertical stripe in the image, as shown in Figure 2.22a. Smear can appear if the V-CCD receives photons due to poor light-shield performance, or if the generated electrons diffuse into the V-CCD. Smear is prevented by placing a light shield close to the photodiode with a specialized CCD process. A low smear-to-signal ratio of −100 dB has been achieved with the IT-CCD image sensor.

Blooming is a phenomenon where excess charge (charge greater than the full-well capacity of the photodiode) spreads to neighboring pixels, and is seen as a white spot/a white vertical stripe, as shown in Figure 2.22b. The VOD structure (see Section 2.1 and Figures 2.5 and 2.6) suppresses blooming.

2.2.5.2 Array Performance

a. Shading

Shading is a slowly varying or low spatial frequency output variation seen in a reproduced image. There are two types of shading: optical shading and electrical shading.

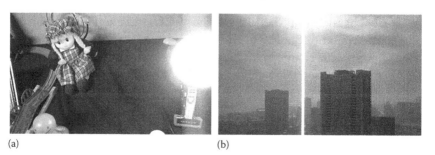

(a) (b)

FIGURE 2.22 (a) Smear and (b) blooming in the CCD image sensor.

The causes of optical shading are the imaging lens itself and the relationship between the angle of incidence of the light and the upper structure of the photodiode (on-chip microlens, CFA, and metal routing), which will be addressed in the following section. With regard to the lens, even if it is perfect, there is a physical limitation, called the cosine fourth law, which states that illuminance declines the further one gets from the center of the image. The cosine fourth law is given as

$$I = I_0 \cdot \cos \theta^4 \qquad (2.14)$$

where I is the illuminance on the focal plane and θ is the incident light angle with respect to the axis of the imaging lens.

Regarding electrical shading, in the IT-CCD image sensor, shading could occur if the resistance on the V-CCD driving lines is large, which results in a degradation of the waveforms of the V-CCD pulse trains. This is what happens when the V-CCD pulse trains are carried along poly Si wirings. To reduce the wiring resistance, a shunt wiring technique is used where metal lines (with a much lower resistance than poly Si) are used to shunt poly Si V-CCD pulse lines.

Also, in the IT-CCD, vertical shading caused by a dark current in the V-CCD register could appear since the charge transfer period for a packet of signal electrons along the V-CCD register varies depending on the vertical position of the pixel (see Figure 2.10a).

In the CMOS image sensor, potential gradients on power and ground buses in the peripheral circuitry could cause shading. Careful design considerations are needed to suppress electrical shading.

b. Angular response and F-number dependence

As shown in Figure 2.23, the angle of incidence of the light from the imaging lens with respect to the on-chip microlens of the image sensor varies depending on the position of the pixel. A portion of the incident light in the array periphery is blocked by the light-shield metal and could be absorbed in the Si bulk outside the pixel, which results in decreased sensor output. An example of the angular dependence is shown in Figure 2.24a.

The angular dependence directly influences the relationship between the sensor output levels and the lens *F*-number, as shown

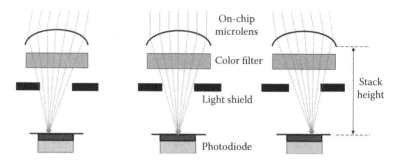

FIGURE 2.23 Mechanism of angular response. (*left*) Leftmost array periphery; (*center*) array center; (*right*) rightmost array periphery.

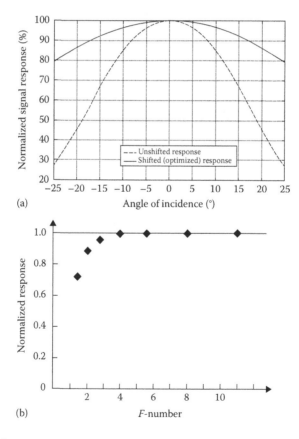

FIGURE 2.24 Example of (a) angular response and (b) *F*-number dependence.

in Figure 2.24b, since the maximum incident light angle varies with the lens F-number (lower F-numbers contain light rays with larger angles of incidence). To improve the angular response and the F-number dependency, a technique to shift microlens positions is commonly employed.

The CCD image sensor exhibits better angular response and F-number dependence than the CMOS image sensor because the stack height is lower in the CCD image sensor. Although efforts have been made to reduce the stack height in the CMOS image sensor, this is a serious issue in small pixel CMOS image sensors since the ratio of the stack height to the pixel size increases as the pixel size decreases. As previously explained, the light guide in the FSI device (see Section 2.4.5) and BSI sensors (see Section 2.4.4) have been developed to help reduce the effective stack height.

c. Column FPN

Suppressing the structural noise is one of the critical requirements for high-quality images in the CMOS image sensor. FPN caused by the threshold voltage variation of the in-pixel source follower transistor would be seen as a spatially random FPN that can be suppressed by the on-chip FPN suppression circuit, as described in Section 2.2. Column FPN, which is a type of structural noise and is seen as vertical stripes in a reproduced image and thus is more noticeable than the spatially random FPN, is caused by a variation in the performance of the column circuitry. Figure 2.25 shows an example of dark images with column FPN (the image is

FIGURE 2.25 Column FPN in the CMOS image sensor.

enhanced digitally to show the column FPN clearly). The visibility of the structural noise can be measured by the ratio of its amplitude to the background random (temporal) noise and it depends on a camera tone curve. To suppress the column FPN, digital CDS (Yang et al. 1999) is commonly employed in the digital-output CMOS image sensors, where two A/D conversions, one for the "reset" level (with zero signal charge) and the other for the "signal" level (with the signal charge) of the pixel output signal, are performed and subtracted from each other.

Obviously, such a circuit-oriented column FPN does not exist in the CCD image. However, if defects that deteriorate the charge transfer efficiency or that generate large dark current variations exist in the V-CCD register, they produce column FPN.

d. Row temporal noise

Row temporal noise is another significant structural noise component of the CMOS image sensor. Since all column circuits operate in parallel, noise injected at a critical time of operation affects the signals of all columns simultaneously, resulting in horizontally correlated noise, called row temporal noise.

It is possible to avoid row temporal noise by setting the S/H timing properly and/or by separating power supplies for the column blocks. Figure 2.26 shows an example of row temporal noise (the image is enhanced digitally to show the row temporal noise clearly).

FIGURE 2.26 Row temporal noise.

FIGURE 2.27 A reproduced image from a CMOS image sensor with successful column FPN/row temporal noise suppression: with 8× analog gain and 32× digital gain.

An example of successful on-chip FPN suppression with the digital CDS technique and with row temporal noise suppression is shown in Figure 2.27 (Matsuo et al. 2008).

2.3 Challenges

There are numerous challenges left for designers, such as maintaining performance while reducing the size of pixels, extending the dynamic range of the sensor, and increasing the readout speed of the sensor.

Efforts to reduce the size and increase the number of pixels are ongoing. To maintain sensor sensitivity even as the pixel size decreases, a scaled performance where responsivity in e⁻/lx-s and noise floor in e⁻ normalized by the pixel size remain the same as those of the previous generation sensor, is at least needed.

Extending the dynamic range of sensors is another important challenge and several methods have been proposed and developed. Extending the dynamic range of a sensor makes it possible for details in both the shaded and the highlighted areas of the image to be visible.

If the pixel size is decreased, the number of pixels in any given optical format will increase. Since this increases the amount of information, it is more difficult to read out all of the data at the same frame rate as is currently being used, or at even higher frame rates. Therefore, further enhancement of high-speed readout capability is an important technical challenge. High-speed readout will contribute to reducing rolling shutter artifacts (see Section 2.3.1). In addition, it will make it possible to perform high-speed multiple readouts and/or multiple pixel data sampling to further reduce temporal noise.

A global shutter can eliminate the artifacts of the rolling shutter and several types of CMOS global shutter pixels have been developed and are in use (Figure 2.11a shows an image taken with a CMOS image sensor with a global shutter). However, it has been a challenge to produce a global shutter pixel with a performance comparable to that of the current rolling shutter pixel. There are several difficulties in making such a global shutter: it requires an analog memory inside each pixel, a global shutter pixel tends to be bigger and has lower responsivity and lower full-well capacity, and higher noise and undesirable artifacts associated with the in-pixel analog memory. Achieving a global shutter CMOS sensor that satisfies all these requirements will be a challenge.

In another vein, recently, image sensors with embedded special pixels have been developed, for example, with pixels that sense depth (Kim W. et al. 2012; Kim S.-J. et al. 2012) and with pixels that are used for autofocusing (Uchiyama 2012). This could be a trend of future image sensors.

Other emerging sensor architectures to be employed in computational photography (see, e.g., http://web.media.mit.edu/~raskar/photo/) include multiple subarrays in a single sensor, where each subarray has its own on-chip lens, and multiple subarrays where each subarray has its own imaging lens (an array image sensor). In both of these cases, each subarray sees a slightly different scene so depth information can be extracted.

References

Agranov, G., Smith, S., Mauritzson, R., Chieh, S., Boettger, U., Li, X., Fan, X., et al. (2011), Pixel continues to shrink ... Small pixels for novel CMOS image sensors, *Proceedings of the International Image Sensor Workshop*, 1–4.

Amelio, G. F. (1973), Physics and applications of charge-coupled devices, *IEEE Intercon Technical Papers*, 1–6.

Bayer, B. E. (1976), Color imaging array, U.S. patent 3,971,065.

Boyle, W. S. and Smith, G. E. (1970), Charge coupled semiconductor devices, *The Bell Systems Technical Journal*, 49, 587–593.

Fossum, E. R. (1993), Active pixel sensors: Are CCDs dinosaurs?, in Charge-Coupled Devices and Solid-State Optical Sensors III, *Proceedings of SPIE, 1900*, eds E. M. Granger and K. T. Knox, SPIE, Bellingham, WA, pp. 2 – 14.

Fossum, E. R. (1997), CMOS image sensors: Electronic camera-on-a-chip, *IEEE Transactions on Electron Devices*, *44*(10), 1689–1698.

Fossum, E. R. (2012), Quanta image sensor: Possible paradigm shift for the future. Paper presented at Grand Keynote, IntertechPira Image Sensors 2012, London, March 22, http://ericfossum.com/Presentations/2012%20March%20QIS%20London.pdf.

Ishihara, Y. and Tanigaki, K. (1983), A high photosensitivity IL-CCD image sensor with monolithic resin lens array, *IEEE Electron Devices Meeting*, *29*, 497–500.

Ishihara, Y., Oda, E., Tanigawa, H., Teranishi, N., Takeuchi, E., Akiyama, I., Arai, K., Nishimura, M., and Kamata, T. (1982), Interline CCD image sensor with an anti blooming structure, *IEEE Solid-State Circuits Conference. Digest of Technical Papers*, *25*, 168–169.

Kim, S.-J., Kang, B., Kim, J., Lee, K., Kim, C-Y., and Kim, K. (2012), A 1920 × 1080 3.65 μm-pixel 2D/3D image sensor with split and binning pixel structure in 0.11 μm standard CMOS, *ISSCC Digest of Technical Papers*, 396–397.

Kim, W., Yibing, W., Ovsiannikov, I., Lee, S., Park, Y., Chung, C., and Fossum, E. (2012), A 1.5 Mpixel RGBZ CMOS image sensor for simultaneous color and range image capture, *ISSCC Digest of Technical Papers*, 392–393.

Kosonocky, W. F. and Carnes, J. E. (1971), Charge-coupled digital circuits, *IEEE Journal of Solid-State Circuits*, *SC-6*(5), 314–322.

Krambeck, R. H., Walden, R. H., and Pickar, K. A. (1971), Implanted-barrier two-phase charge-coupled device, *Applied Physics Letters*, *19*(12), 520–522.

Matsuo, S., Bales, T., Shoda, M., Osawa, S., Almond, B., Mo, Y., Gleason, J., Chow, T., and Takayanagi, I. (2008), A very low column FPN and row temporal noise 8.9M-pixel 60 fps CMOS image sensor with 14bit column parallel SA-ADC, *Symposium on VLSI Circuits Digest of Technical Papers*, 138–139.

Recommendation ITU-R BT.709-5 (2002), Parameter values for the HDTV standards for production and international programme exchange, http://www.itu.int/dms_pubrec/itu-r/rec/bt/R-REC-BT.709-5-200204-I!!PDF-E.pdf.

Takayanagi, I. (2006), CMOS image sensors, in *Image Sensors and Signal Processing for Digital Still Cameras*, ed. J. Nakamura, CRC Press, Boca Raton, FL, pp. 143–177.

Takayanagi, I. and Nakamura, J. (2013), High-resolution CMOS video image sensors, *Proceedings of the IEEE*, *101*(1), 61–73.

Teranishi, N. (2000), Progress in solid-state image sensors, *ITEJ*, *54*(2), 141–147.

Teranishi, N., Kohno, A., Ishihara, Y., Oda, E., and Arai, K. (1982), No image lag photodiode structure in the interline CCD image sensor, *IEDM Technical Digest*, 324–327.

Theuwissen, A. J. P. (1995), *Solid-State Imaging with Charge-Coupled Devices*, Kluwer, Dordrecht.

Theuwissen, A. J. P. (2006), The hole role in solid-state imagers, *IEEE Transactions on Electron Devices*, *53*(12), 2972–2980.

Uchiyama, S. (2012), Superiority of image plane phase detection AF, *ITE Technical Report*, *36*(38), 17.

Wakabayashi, H., Yamaguchi, K., Okano, M., Kuramochi, S., Kumagai, O., Sakane, S., Ito, M., et al. (2010), A 1/2.3-inch 10.3Mpixel 50frame/s back-illuminated CMOS image sensor, *ISSCC Digest of Technical Papers*, 410–411.

Walden, R. H., Krambeck, R. H., Strain, R. J., McKenna, J., Schryer, N. L., and Smith, G. E. (1971), The buried channel charge-coupled devices, *The Bell System Technical Journal*, *51*, 1635–1640.

Weckler, G. P. (1967), Operation of p–n junction photodetectors in a photon flux integration mode, *IEEE Journal of Solid-State Circuits*, *SC-2*(3), 65–73.

White, M. H., Lampe, D. R., Blaha F. C., and Mack, I. A. (1974), Characterization of surface channel CCD arrays at low light levels, *IEEE Journal of Solid-State Circuits*, *SC-9*(1), 1–13.

Yamada, T. (2006), CCD image sensors, in *Image Sensors and Signal Processing for Digital Still Cameras*, ed. J. Nakamura, CRC Press, Boca Raton, FL, pp. 95–141.

Yang, W., Kwon, O.-B., Lee, J.-I., Hwang, G.-T., and Lee, S.-J. (1999), An integrated 800 × 600 CMOs imaging system, *ISSCC Digest of Technical Papers*, 304–305.

CHAPTER **3**

Voice Coil Motors for Mobile Applications

Luke Lu

Contents

3.1 Introduction

Voice coil motors (VCMs), or so-called linear motors, are well-known electromechanical components (Hollerbach et al. 1992). VCMs have been widely used for devices such as computers, automobiles, and monitors (see Figure 3.1). During the last decade, renewed interest was seen in mobile applications to enable autofocus (AF) functionality. Thus, the AF linear motors

83

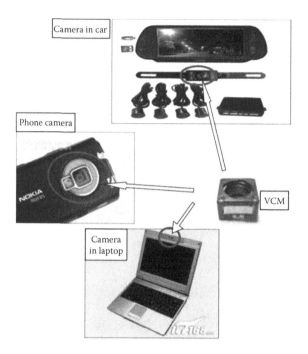

FIGURE 3.1 Examples of multiple applications of the VCM in various mobile cameras.

quickly became the dominating technology used in mobile-phone camera modules.

3.1.1 Basic Optical Considerations When Using VCMs

If we use a camera to take a picture for various object distances (e.g., when the object distance is 5 or 0.1 m), we will find that the corresponding image positions (or distances) are different if we use the same optical lens. Thus, when the object distance S_1 and the focal length f of the lens are fixed, then the image distance S_2 can be calculated, according to the formula for lens imaging (Hecht 2001):

$$\frac{1}{S_2} = \frac{1}{f} - \frac{1}{S_1} \tag{3.1}$$

For example, for a lens with an effective focal length (EFL) of 4.88 mm, if the first object distance is $S_{1,1} = 5$ m, we can calculate the corresponding imaging distance as $S_{2,1} = 4.855$ mm; while if the object distance is $S_{1,2} = 0.1$ m, the imaging distance will be $S_{2,2} = 5.130$ mm. Thus, we find a difference of $S_{2,2} - S_{2,1} = 0.245$ mm

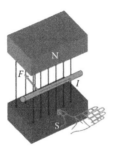

FIGURE 3.2 Schematic demonstration of the directions of the magnetic field (N–S) of the electrical current *I* and of the force applied on the wire by the magnetic field.

in the imaging distance for those two cases. It means that if we have already set the lens and imaging distance for one object (say at 5 m), then to take a picture of another object, which is at 0.1 m, we should shift the lens by 0.245 mm. Therefore, depending on the object distance, we can use the VCM to adjust the focal plane of the lens assembly, in order to optimize the image on the photo-sensitive device (e.g., complementary metal-oxide-semiconductor [CMOS] or charge-coupled device [CCD]; see Chapter 2), which is usually in a fixed position. The resulting imaging, therefore, is often called the "focus module" and the VCM is called the "focus motor."

3.1.2 Basic Physics of VCM Operation

The VCM operation may be explained by using basic electro-magnetic theory. The operation of the VCM benefits from the force on the electrified wire (coil) from a magnetic field (shown in Figure 3.2). This phenomenon has been studied in detail and has largely been used for developing VCM-type devices.

3.1.3 Structure of VCMs

The electrical wire is usually placed inside a uniform, paral-lel magnetic field, which is created by permanent magnets (see Figure 3.3). After being electrified, the wire can be moved by the magnetic field. For a simplified description, we can assume that the direction of the magnetic field is in the drawing plane, as shown in Figure 3.3.

At the same time, we can also assume that the direction of the current in the wire is perpendicular to the plane of the draw-ing, as shown in Figure 3.3. Then, according to electromagnetic theory, the resulting force and displacement of the wire will be in the plane of the drawing, shown by the bold vertical arrow in

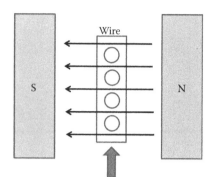

FIGURE 3.3 Schematic presentation of the electrical wire (the current is perpendicular to the plane of drawing, toward the viewer), the magnetic field of the VCM (in the plane of drawing), and the direction of movement (in the plane of drawing, shown by the bold vertical arrow).

Figure 3.3. The structure and the moving direction of a VCM are shown in Figure 3.4.

However, it is important to realize that the electromagnetic force alone is not sufficient to make the wire stop at the required position. Therefore, the spring is added as a reacting force with respect to the electromagnetic force. The spring force balances the electromagnetic force and stops the lens assembly at the correct position. This operation uses the balance of forces, including the electromagnetic force applied on the coil, F_c:

$$F_c = n \cdot BIL_b \qquad (3.2)$$

where
  B is the magnetic field intensity
  I is the loaded current
  L_b is the length of the coil

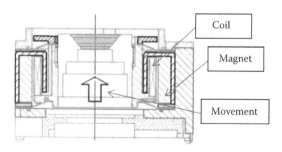

FIGURE 3.4 Structure and movement of the VCM.

n is the number of magnetic field sources (usually $n = 1$)
F_s, the spring force, is

$$F_s = K \cdot \Delta \tag{3.3}$$

where
K is the force constant of the spring
Δ is the amount of elastic deformation of the spring

The displacement function may thus be found as

$$\Delta = \frac{1}{K} \cdot n \cdot BIL_b - f \tag{3.4}$$

where f is the amount of elastic deformation in the original state of the spring.

So, the total VCM stroke equals the total deformation of the spring minus the amount of elastic deformation in the original state.

If we suppose $G = 1/K \cdot n \cdot BL_b$, then the simplified displacement function may be expressed as

$$\Delta = G \cdot I - f \tag{3.5}$$

From Equation 3.5, we can find the displacement of the VCM that is defined by the value of the current (I).

3.1.4 Functions of Different Parts in VCMs

A VCM is commonly composed of about a dozen parts (see Figure 3.5), grouped into four categories: stator (core), rotor,

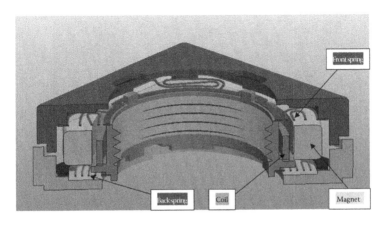

FIGURE 3.5 Cross-sectional view of the main components of the VCM.

TABLE 3.1   Groups, Parts, and Their Functions in a VCM Component

Groups	Parts	Functions
Stator (core)	Yoke	To sustain the weight of the magnets and also to confine one magnetic field region
	Magnet	To provide a stable magnetic field
Rotor	Coil	To conduct electrical current and move
	Carrier	To sustain the weight of the optics and move with the coil
Spring assembly	Front spring	To provide a reacting force to fix the carrier
	Back spring	To provide a reacting force to fix the carrier
Fixing modules	Front spacer	To fix the front spring and give some space for forward moving
	Back spacer	To fix the back spring
	Base	To fix other parts of the VCM, connecting to the other part of the camera module

spring assembly, and fixing modules. The stator (core) is composed of the yoke and the magnet; the rotor is composed of the coil and the carrier; the spring assembly is composed of the front spring and the back spring; and the fixing modules are composed of some plastic spacers (see Table 3.1 and Figure 3.6).

3.2 Technical Evaluation of VCMs

Several important parameters must be checked to evaluate the VCMs. Thus, we need to check the stroke, starting current, hysteresis tolerance, stroke increment, theoretic linear tolerance, and multi form.

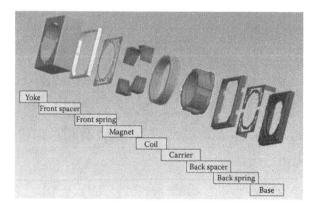

FIGURE 3.6   Exploded view of separate parts of the VCM.

Stroke of VCM: The largest displacement of a VCM, when the maximum value of a current is applied to the VCM (e.g., 80 mA in Figure 3.7). It is used to evaluate whether the amount of displacement of the lens is enough to make a good image. If the value of the lens displacement is greater than the amount that the lens actually needs, then that will cause some blind spots where we will not have a good image. Usually, a VCM with an overall dimension of 8.5 × 8.5 mm needs to have a stroke of about 200–300 μm for use in a 1/4-inch sensor 5M camera.

Starting current: The minimum current value that is necessary for the VCM to start to move. The VCM will start to move if the stroke is above the established value (e.g., when the VCM's stroke is 10 μm), as shown in Figure 3.7. This feature is used to evaluate the stability of the lens system and the springs, under the initial movement. If the start current is too small, the lens system will not be stable enough and it will not be able to focus accurately and stably. A starting current that is too large would be a waste of the electrical energy. On average, the starting current may be in the range between 15 and 30 mA or between 25 and 50 mA, according to the customer's demands.

Hysteresis tolerance: Hysteresis is present in almost all mechanical moving systems. It is the displacement difference between the VCM moving forward and backward under the same current

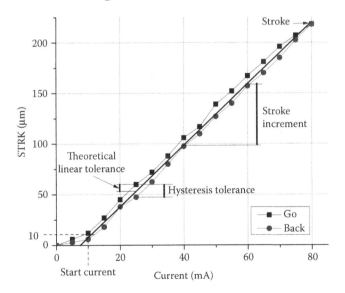

FIGURE 3.7 Demonstration of various key parameters in the dependence of the stroke versus the driving current.

value. It shows whether the VCM will return to the same position after it moves (see Figure 3.7). Hysteresis tolerance (typically ±10 μm) is used to evaluate the consistency of the VCM during focusing. If the VCM has a good consistency, when the focus is complete, it will return close to the required position.

Stroke increment: This is the displacement of the VCM per milliampere (mA) of current, which manifests the size of the step of the VCM. In Figure 3.7, the steady region between 40 and 60 mA is usually selected for calculating the stroke increment, with the formula: Stroke increment = (stroke 60 mA − stroke 40 mA)/20. If the stroke increment is too small, the maximum stroke cannot reach the required position at 80 mA (note that the maximum output current may be of the order of 100 mA [Texas Instruments, Voice Coil Motor Driver (DRV201) for Camera Auto Focus, www.ti.com]).

Theoretic linear tolerance: The difference between the actual displacement of the VCM and the displacement predicted by theory. It is used to check whether the VCM walks straight (see the deviation from the straight line in Figure 3.7).

Multi form: This is another way to evaluate the theoretic linear tolerance. It is used to verify whether "the curve of VCM displacement versus current" has one or more slopes (Figure 3.8).

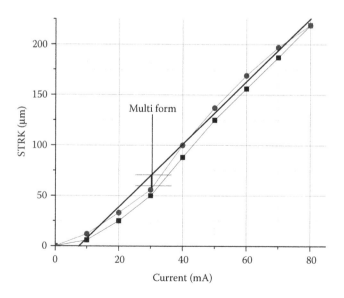

FIGURE 3.8 Multi form.

Only once those key parameters are characterized will the VCM be considered for the AF application. However, there are still some additional parameters that must be taken into account.

3.2.1 Effect of Gravity

In a large majority of cases, pictures are taken in the horizontal direction. However, if we want to take pictures upward or downward (e.g., for bar coding or text scanning), we must consider the gravity of the lens system. The gravity of the lens, g, will add to the spring force, F_s. Thus, when the VCM moves upward, the electromagnetic force is supposed to balance the force $F_s + g$. In contrast, when the VCM moves downward, the electromagnetic force should balance the force $F_s - g$, the normal tolerance to the corresponding positional difference is less than ± 50 μm (see Figure 3.9).

3.2.2 Displacement Testing (Stroke Test)

The displacements of a VCM (typically up to 0.2–0.25 mm) corresponding to different current values are measured using displacement testing, which is a basic test for cameras. Generally, stroke tests are used for displacement testing in the 5M camera modules. The stroke test requires very precise instruments. Nowadays, laser range finders are used to do this measurement. To perform the test

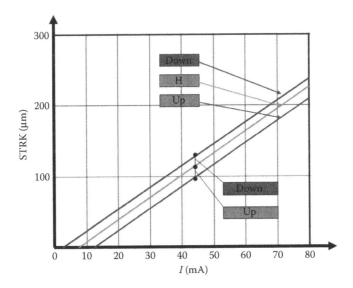

FIGURE 3.9   Schematic demonstration of the gravity effect for camera orientations pointing down, horizontally (H), and up.

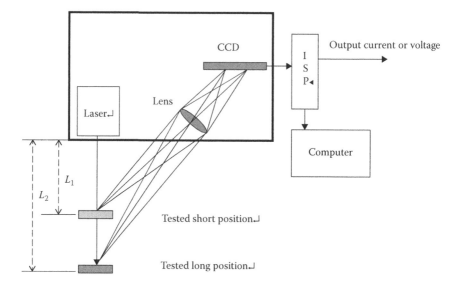

FIGURE 3.10 Schematic demonstration of the experimental setup using a laser range finder.

and obtain the results, described in Figure 3.10, we first align the detection region of the laser range finder, record the position of the VCM when no current has been applied, then we set the displacement value to zero. Subsequently, the current is varied from 0 to 80 mA dc, while we measure the corresponding position of the VCM. In this way, we can plot the VCM movement graph, from which we can judge whether the VCM is acceptable.

The principle of such a measurement is the triangulation method (see Figure 3.10). Assuming that the object moves from L_1 to L_2, then its imaging on a CCD camera will change correspondingly. So, by measuring the position shifting of the imaging on the CCD, we can measure the displacement of the object. The maximum displacement is ABS $(L_1 - L_2)$.

3.2.3 Optical Axis Tilting

The test of the optical axis is one of the most essential requirements for a high-quality camera module. In the VCM, the lens is fixed by the spring assembly. The structure of the spring assembly is shown in Figure 3.11. Due to the structure and the inaccuracy of a spring, the force from the springs at four corners cannot be exactly uniform. In this case, the optical axis of the lens assembly will tilt (see Figure 3.12), for example, up to 15′. Thus, the optical axis testing is mainly done to see whether the lens tilts when the VCM is set still or when the lens moves upward.

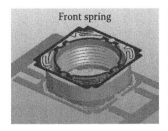

FIGURE 3.11 Demonstration of front and back spring assembly.

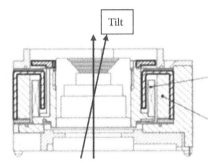

FIGURE 3.12 Schematic demonstration of the tilt of the optical axis of the VCM.

The test method is similar to the displacement test. Generally, it is divided into:

- Initial optical axis (static tilt): the angle deviation between the zero point and the initial point
- Changing optical axis (dynamic tilt): the maximum value of the deviation of the initial point during the action
- Integrated optical axis (total tilt): the maximum value of the angular deviation of the zero point during the whole moved action

3.2.4 Step Response of Spring Assembly

In a VCM, two springs can be used to hold the fixed lens carrier. When we add the applied force (F) to the spring, it may be extended up to the maximum stroke. The spring will correspondingly develop a reactive force (F_s) against the stroke. By inertia, the needed equilibrium position will not be achieved in a monotonic way and the jitter will need some time to achieve a stable state, as shown in Figure 3.13.

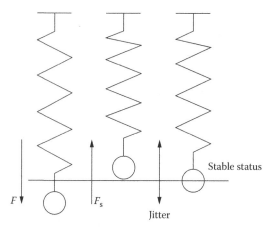

FIGURE 3.13   Schematic demonstration of the jitter effect.

Thus, since the lens in the VCM is fixed by two springs, when we simply add a current from 0 to 20 mA, then relatively long damping oscillations occur (because of the spring jitter or the so-called ringing effect) before its stabilization. We call this time the stabilization time, shown in Figure 3.14. To optimize the AF convergence time, it is necessary to minimize the VCM stabilization time. Specific driving algorithms are developed (Liu et al. 2009) to accelerate this transition and reduce the ringing effect (e.g., by

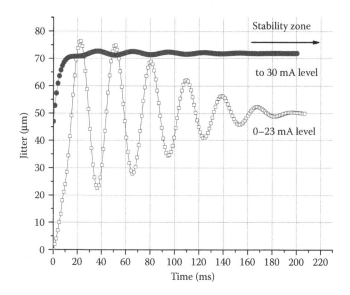

FIGURE 3.14   Jitter effect and stabilization time of the VCM.

passing to 30 mA, see Figure 3.14). Then the transition may take in the order of a few milliseconds up to 20 msec. Obviously, the FD search algorithms must then be adapted to take into account this phenomenon and to carefully synchronize those transitions with the image frames of the camera (Hsu and Fuh 2006).

3.2.5 Reliability Testing

In addition to the above-mentioned performance tests, the VCMs still need to be tested for their reliability aspects, which includes testing the drop, durability, vibration, high temperature, low temperature, high humidity, vibration of transport packaging and simulation of the actual environment, and storage. The typical duration of the quality guaranty of the VCM is approximately 6 months after the product delivery (at ambient temperature and humidity; a temperature between 15°C and 35°C and a relative humidity [RH] between 30% and 80%). Within this period of time, VCMs (typically a couple of tens of units) must pass (maintain their main specifications) tests such as

- High-temperature storage: +85°C ± 2°C for 500 h
- Low-temperature storage: −40°C ± 2°C for 500 h
- Operating temperature/humidity: +80°C ± 2°C and a RH of 95% for 500 h of operation (0–80 mA/30 sec: 0 mA~15 sec, 80 mA~15 sec)
- Temperature/humidity storage: +80°C ± 2°C and a RH of 90% for 500 h
- Thermal shock test: +85°C ± 5°C/30 min; ~+40°C ± 5°C/30 min; ambient: 3 min; for 300 cycles
- Life test: 0~100~0 mA for 1 sec; 300,000 cycles of 0.5 sec ON and 0.5 sec OFF for +80°C ± 2°C, +25°C ± 2°C, and −30°C ± 2°C
- Shock test (with lens unit): acceleration 1000 m/sec², pulse width 6 msec, and frequency of shock three times each; all axes
- Drop test (with lens unit): from 106 cm to the concrete floor (fixture weight 160 g), six directions, three times per direction
- Vibration test: three axes, 30 min per axis, frequencies ranging from 5 to 500 Hz (component operating)

These conditions may change for various suppliers. All these elements are potential threats and the VCM must pass these tests to make sure that the product will maintain a good performance.

3.3 Future Trends

VCMs are under continuous pressure for further optimization (Liu and Lin 2008; Chung 2005). New designing tools are currently used to make their development process more effective. Thus, in addition to the traditional mechanical tools, new analytical tools and software are being developed to perform the simulation of the electromagnetic induction, as shown in Figure 3.15.

These simulation tools are also used to calculate the spring in the modulus of elasticity and the amount of elastic deformation.

Further effort must be made to reduce the power consumption of a VCM for mobile AF applications, which may be of the order of 250 mW. Finally, and perhaps most importantly, VCMs are continuously threatened by new emerging technologies and their price must be reduced to maintain their dominant position in the market.

3.4 Other Possible Applications

Currently, the VCM's maturity is high and various other applications are already starting to penetrate the market, for example, optical image stabilization (OIS) and optical zoom (OZ). The OIS application requires a rather complex system with multiple VCMs, which increases its corresponding size, power consumption, fragility, and price. However, the first VCM-OIS products are already available in the market since 2012. The OZ requires even more volume and thus it is currently less attractive for mobile (miniature) cameras.

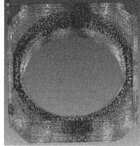

FIGURE 3.15 Demonstration of a simulation of the VCM (*left*) and its performance (*right*).

3.5 Summary and Conclusions

Today, VCMs are largely used in various consumer electronic products. They are dominating the AF application market in mobile phones. Several new technologies, such as ultrasonic motors and liquid crystal, are aspiring to replace VCMs or at least some specific segments of the market. However, VCMs are the milestone product in this phase of microcamera history.

Acknowledgment

I would like to thank Professor Galstian for his help during the work on this chapter.

References

Chung, M.J. (2005), Development of compact auto focus actuator for camera phone by applying new electromagnetic configuration, *Proceedings of SPIE, the International Society for Optical Engineering, 6048*, 60480.

Hecht, E. (2001), *Optics* (4th edn), Addison-Wesley, Reading, MA.

Hollerbach, J., Hunter, I., and Ballantyne J. (1992), A comparative analysis of actuator technologies for robotics, in O. Khatib, J. Craig, and R. Lozano-Perez (Eds), *The Robotics Review 2*, MIT Press, Cambridge, MA, pp. 299–342.

Hsu, W. and Fuh, C.S. (2006), A new sampling method of auto focus for voice coil motor in camera modules, in L.-W. Chang, W.-N. Lie, and R. Chiang (Eds), *PSIVT 2006, LNCS 4319*, Springer-Verlag, Berlin, pp. 1254–1263.

Liu, C.S. and Lin, P.D (2008), A miniaturized low-power VCM actuator for auto-focusing applications, *Optics Express 16*(4), 2533.

Liu, C.S., Lin, P.D., Lin, P.H., Ke, S.S., Chang, Y.H., and Horng, J.B. (2009), Design and characterization of miniature auto-focusing voice coil motor actuator for cell phone camera applications, *IEEE Transactions on Magnetics 45*(1), 155.

CHAPTER 4

Extended Depth of Field Technology in Camera Systems

Dmitry Bakin

Contents

4.1 What Is Extended Depth of Field Technology?

In a standard focused camera system, the lens forms a sharp image of a remote object on the surface of the sensor array. When the object gets closer to the camera, the incident wave front becomes curved and exhibits phase delay at the lens aperture. The optical wave front is now converging on a plane shifted backward from the original focal plane. As a result, the image gets defocused, and loses its sharpness. That loss of sharpness becomes very drastic with modern miniature camera systems as the pixel size is reduced and the sensor format size is increased. The range of object distances within which the image is perceived as sharp by an observer is called the depth of field (DOF) of an optical system.

To overcome the problem with image defocusing, standard high-resolution camera systems usually employ various "active" autofocusing (AF) elements, which allow a shifting image focal plane and bring it back to focus on the sensor array surface. The AF systems provide an efficient and natural way of creating a sharp image of the object as perceived by the observer, but they usually create delays in the operation of cameras as well as increases in camera power consumption, camera size, and system cost. On the other hand, several alternative fixed-focus "passive" methods of extending the DOF of the optical system are also implemented with miniature cameras. They do not require AF technology and instead are based on combining apodized camera optics with digital image-processing algorithms. Extended depth of field (EDOF) technology incorporates modified optics and digital processing for the simultaneous in-focus imaging of near and faraway objects.

The driving force behind the development of EDOF technology is the modern trend for miniaturizing complementary metal-oxide-semiconductor (CMOS) sensors and imaging optics, and integrating with powerful digital processing. Miniature cameras incorporating high-resolution sensors and fixed-focus optics allow close-range reading of printed material (bar-code patterns and business cards), while providing high-quality imaging in more traditional applications. These cameras incorporate modified optics and digital processing to recover the soft-focus images and restore sharpness over a wide range of object distances. Various image-processing methods are used for restoring the image (Jorge and Berriel-Valdos 1990; Grossmann 1987; Dowski and Cathey 1995). In general, the lens system is modified to introduce wave-phase variations into the optical path, so that the point spread function (PSF) of the lens become larger, but stays well confined, monotonic, and invariant over the extended range of object distances (Chi and George 2003). Prior knowledge of the PSF distribution allows the application of advanced deconvolution methods to recover the image of the scene with the objects staying in focus over the design distance range.

4.1.1 Main Advantages and Applications of EDOF Technology

Typically, the focus recovery (FR) procedure involves tradeoffs between image quality attributes, such as DOF, sharpness, visual noise, color reproduction, and edge artifacts. EDOF technology introduces a new set of tradeoff parameters achievable for camera image processing (Figure 4.1).

The passive EDOF systems opened the way for the introduction of high-resolution cameras at low cost and in smaller packages for a variety of mobile imaging applications (low z-height cameras, bar-code readers, business-card readers, etc.). Also, as shown in Figure 4.2, in EDOF technology, the nature of the field

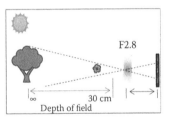

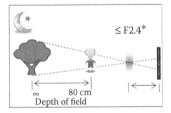

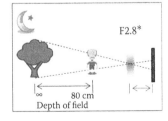

FIGURE 4.1 Main advantages for camera systems with EDOF technology.

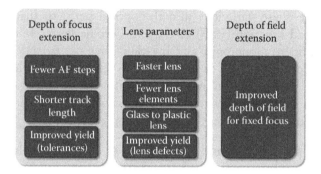

FIGURE 4.2 Areas of performance improvement and cost reduction.

depth extension also allows for creating systems with fewer AF steps, and more relaxed optical alignment tolerance. Additional cost reduction and performance improvement tradeoffs could be realized by using a faster lens or fewer lens elements.

The DOF in a high-resolution camera is related to various parameters of the imaging module, such as the pixel size, resolution, lens focal distance, and aperture stop. The limits on the achievable EDOF range are governed by the optical properties of the lens, the pixel characteristics of the sensor, and the acceptable signal-to-noise ratio (SNR) in the output image after applying the restoration filter. These parameters define the boundaries of the DOF extension factors when applied to high-resolution CMOS modules. The extension factors for the EDOF imaging module are defined in terms of an improved absolute resolution in object space while maintaining focus at infinity. This definition is used for the purpose of identifying the minimally resolvable object details in mobile cameras with a bar-code reading feature. The performance of miniature cameras can also be analyzed and compared in terms of the quality of pictorial imaging. The method of comparing pictorial image resolution based on the subjective quality factor (SQF) (Granger and Cupery 1972) allows for a through-focus analysis of EDOF extension factors.

4.2 Generic Optical Properties of EDOF Cameras

A typical fixed-focus imaging module lens is aligned and secured within the mounting barrel to provide a sharp image of distant objects on a photosensitive pixel array sensor. When an object gets closer to the camera, the lens forms an image that moves behind the sensor array plane.

The convenient way to define the DOF for an imaging system is to use the Newtonian form of the basic lens equation (Equation 4.1). The conjugated object and the image planes of the lens are related in terms of distances x_1 and x_2 as measured from the front focus (FF) and the back focus (BF) planes of the imaging system, as shown in Figure 4.3:

$$x_1 \times x_2 = F^2 \tag{4.1}$$

When an object is at infinity ($x_1 = \infty$), the image is exactly at the BF position ($x_2 = 0$). When an object moves closer, the image moves away from the focus, so that

$$x_2 = \frac{F^2}{x_1} \tag{4.2}$$

When an ideal lens forms an image behind the sensor array plane, the marginal rays spread out over several pixels creating a "blurred" effect on the sensor. This effect is manifested as an increase of the PSF of the optical system, with a corresponding decrease in the resolution and the modulation transfer function (MTF).

Figure 4.4 illustrates the process of spreading a point image spot into neighboring pixels for an ideal lens, when the image is formed beyond the plane of a sensor array. The pixel size is characterized by the effective size a_e. As the blur of an imaging spot δ spreads beyond a single effective pixel of a sensor array, that is perceived by an observer as a "just noticeable difference" (JND) in the softening of an image. As such, the object distance at which

$$\delta \geq a_e \tag{4.3}$$

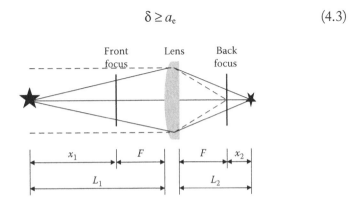

FIGURE 4.3 Positions of lens conjugate points (notice that rays from infinity are converging at BF).

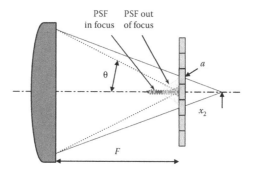

FIGURE 4.4 PSF spread of a defocused image beyond the effective pixel of a sensor array creates a "blur" effect.

defines the so-called hyperfocal distance (HFD) of a camera. Using Equation 4.2 for defocus in distance x_2, and expressing it in terms of pixel size and lens aperture stop F-number ($f\#$) $x_2 = a_e f\#$, we get the expression for HFD as:

$$\text{HFD} = \frac{F^2}{a_e f \#} \qquad (4.4)$$

The conjugate distance in the image plane corresponding to 1 HFD determines the DOF of the camera. This is an axial shift in the image plane characterized by the appearance of 1 JND in an image blur.

4.2.1 EDOF Range for Imaging Module

The effective pixel size of a monochrome sensor is equal to the actual pixel size of an array, while in a color sensor with a Bayer color filter array (CFA) pattern the effective pixel size equals twice the pixel size for red and blue colors and ×1.41 of the pixel size for green.

Consider an example of an imaging module with an effective pixel size $a_e = 7.2$ μm and an ideal diffraction-limited lens defined by the focal length $F = 2.5$ mm, and the aperture stop $f\# = 2.8$. According to formulas 4.4 and 4.1, this module would have a HFD = 310 mm, and a DOF = ± 20 μm. When the lens is best-focused for an object position at infinity, the closer objects will appear in-focus as long as their images are not shifted beyond the sensor array by more than 20 μm.

The operational focus-free range (FFR), without AF, would cover distances from infinity to 310 mm. The graphic in Figure 4.5

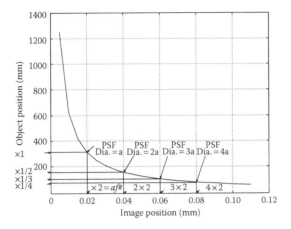

FIGURE 4.5 Relation between object and image positions in a camera with $F = 2.5$ mm, $a = 7.2$ μm, and $f\# = 2.8$. The best focus is at infinity.

shows the image shift for this module as the function of object distance. The image spot size, when judged by the camera output, increases by a quantum step scaled by the effective pixel size. It follows that the described camera could provide sharp images of objects without focus adjustment for a distance ranging between infinity and 310 mm. At closer distances the image gets out of focus. Since the camera HFD is proportional to the square of the lens focal distance, it would be advantageous for the purpose of focus-free imaging to use a lens assembly with a shorter focal distance F. Of course, that would also mean a larger field of view (FOV) and a less detailed resolution in images of faraway scenes.

For optimal utilization of the camera's DOF, typically the lens is initially focused on the object at HFD. That allows achieving an EDOF range from infinity to HFD/2 without a noticeable sacrifice in the image quality. The graphs in Figure 4.6 show examples of the EDOF range coverage for a typical compact imaging cameras used in mobile phones. Examples are shown for a 3 megapixel (MP) camera with a full diagonal FOV of 60°. The example shown on the left curve is for a sensor with a pixel size of 1.75 μm, and the example shown on the right is for another 3 MP sensor with a pixel size of 2.2 μm. On the graphs, the upper and lower lines designate the limits of the object distances for which the image is perceived as staying in-focus. The central line represents the actual object position. The camera with the larger pixel size has a longer focal length and a smaller EDOF range.

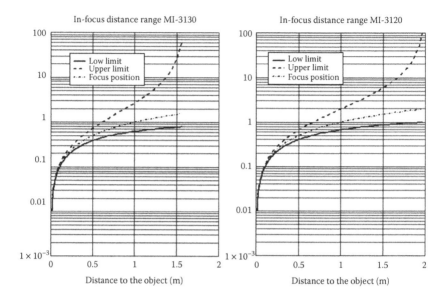

FIGURE 4.6 EDOF range for 3 MP mobile-phone cameras with a diagonal FOV of 60° and a lens aperture stop of f#2.8: left curve pixel size is 1.75 μm and right curve pixel size is 2.2 μm.

A convenient way to analyze the EDOF range of the imaging system is to use units of reciprocal object distances with the factor 1/HFD as the scaling factor (Bakin and Keelan 2008). When the ideal lens is nominally focused for an object at a distance x_0 in front from the camera, the range of perceived "in-focus" distances extends from the minimal distance:

$$x_{1min} = \text{HFD} \times \frac{x_0}{\left(\text{HFD} + x_0\right)} \qquad (4.5)$$

to the maximum:

$$x_{1max} = \text{HFD} \times \frac{x_0}{\left(\text{HFD} - x_0\right)} \qquad (4.6)$$

In units of reciprocal object distances, the EDOF range of the camera is

$$\frac{1}{x_{1min}} - \frac{1}{x_{1max}} = \frac{2}{\text{HFD}} \qquad (4.7)$$

Consequently, the EDOF range for any imaging module expressed in units of inverse HFD will always be equal to 2, and independent of the nominal best focus position x_0 of the lens as well as the object position x_1.

This provides a convenient normalization scale for comparing the EDOF extension factor for different imaging modules, since any conventional diffraction-limited lens will always have an EDOF value of 2. At the same time, the size of a defocused spot δ will depend on the nominal focus position as well as that of an object:

$$\delta = F^2 \times \frac{\left|1/x_1 - 1/x_0\right|}{f\#} \tag{4.8}$$

4.2.2 HFD in Pictorial Imaging

In high-quality pictorial imaging, an "excessive" number of pixels in the image sometimes makes individual pixels "indistinguishable" when viewed under typical observation conditions (at a distance of two picture heights). In this case, the sharpness of the picture is determined based on the perception of the human visual system. By analogy with photography, the term *circle of confusion* (COC) is used to describe the radius of the minimally resolvable spot in the picture when observed under typical conditions. For the 35-mm format, the nearly universally adopted value of the COC is 33 μm, which is about 1/1300 of the format diagonal. Converting this analogy to the digital camera format having an Nh number of horizontal pixels and an Nv number of vertical pixels, the COC will be equal to

$$\delta = \frac{a \cdot \sqrt{Nh^2 + Nv^2}}{1300} \tag{4.9}$$

From Equation 4.9, it follows that in digital photography, pictures made with cameras having 4:3 aspect ratio Bayer sensors, the detectable blur, under typical observation conditions, is pixel size limited until $Nh < 1480$. This transition point corresponds to a sensor having approximately two million pixels (2 MP). When the picture has over 2 MP, the COC would grow in proportion to the square root of the total number of pixels. As a result, the perceived HFD in a pictorial image (HFDc) would grow as the linear function of *Nh*. At the same time, the HFD defined by the effective pixel size (HFDs) would grow as the square of the number of horizontal pixels. The crossing point is around 1480 pixel counts. The described behaviors for the COC and HFD dependencies are illustrated in Figure 4.7 for the sensor with a 4:3 aspect ratio, a pixel size of 1 μm, and a Bayer pattern CFA. In Figure 4.8 (*top*), the horizontal dashed line corresponds to the detectable blur of the point source image when limited by the effective pixel size (example for $a_e = 1.41$ μm), and the solid line corresponds to the scaling-up of COC with an increase in the image resolution as a result of human perception. The crossing point is reached at approximately $Nh = 1480$. In Figure 4.8 (*bottom*) the dashed line again represents the dependence of the HFD from the horizontal pixel count when computed based on the effective pixel size limit HFDs ($a_e = 1.41$ μm), and the solid line is based on the human perception HFDc. In pictorial imaging, as the sensor pixel count increases, the perceived HFD will follow the dashed curve until the sensor pixel count is below 2 MP, and then the solid curve at the higher pixel counts. The specific HFD values for different sensors used in CMOS cameras are presented in Table 4.1.

By comparing the HFD computed based on the effective pixel size (HFDs) and the one based on the COC in human perception (HFDc), one can see that HFDs become dominating over the HFDc when the sensor pixel count exceeds 2 MP, which is in good agreement with the graphical interpretation in Figure 4.8.

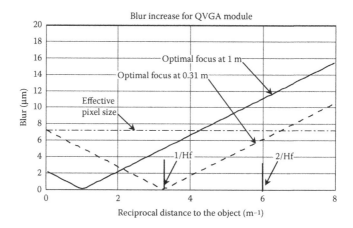

FIGURE 4.7 Illustration of an EDOF range for a QVGA module with $F = 2.5$ mm, $a = 7.2$ μm, and $f\# = 2.8$ (HFD = 310 mm).

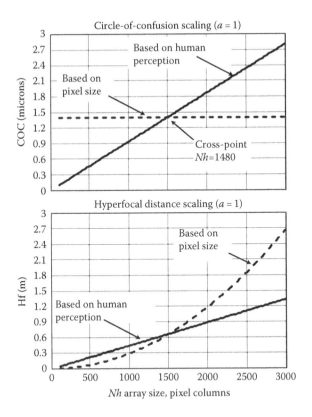

FIGURE 4.8 Computation of the DOF in pictorial imaging for a color sensor: (*top*) circle-of-confusion and (*bottom*) hyperfocal distance (*Nh*: number of pixels in the horizontal direction, pixel size $a = 1$ μm).

4.2.3 Resolution of EDOF Camera in Object Space

One of the emerging applications for handset cameras is the reading of small-print text materials, for example, business cards, bar codes, indoor signs, and regular books and newspapers. The versatile optical character recognition (OCR) software in mobile imaging devices could open the door for new uses in advertising, comparison shopping, traveling, entertainment, etc. Devices with OCR capabilities are expected to become mainstream in the near future. In these applications, the fixed-focus cameras have traditionally had a disadvantage due to loss of image resolution for objects at close distances and hence their inability to resolve small-print texts. The EDOF technology allows this shortcoming to be remedied by restoring the sharpness of out-of-focus objects.

Several factors affect the resolvable power of EDOF cameras in object space: it is limited by the SNR of the image, the recoverable

TABLE 4.1 HFD Values for Mobile Cameras with Different Sensor Resolutions and Pixel Sizes

Sensor Type/ (Format)	Array Columns (Nh)	Array Rows (Nv)	Pixel (μm)	Full Field (°)	Lens Aperture (f#)	Sensor Hyperfocal Distance (HFDs) (m)	Circle-of-Confusion Hyperfocal Distance (m)
VGA/(1/11")	640	480	2.2	60	2.8	0.27	0.62
VGA/(1/6")	640	480	3.6	60	2.8	0.44	1.02
VGA/(1/4")	640	480	5.6	60	2.8	0.69	1.59
1.3M/(1/6")	1280	1024	1.75	60	2.8	0.86	0.99
1.3M/(1/5")	1280	1024	2.2	60	2.8	1.08	1.25
1.3M/(1/4")	1280	1024	2.8	60	2.8	1.37	1.59
1.3M/(1/3")	1280	1024	3.6	60	2.8	1.77	2.04
2M/(1/5")	1600	1200	1.75	60	2.8	1.34	1.24
2M/(1/4")	1600	1200	2.2	60	2.8	1.69	1.56
2M/(1/3.2")	1600	1200	2.8	60	2.8	2.15	1.98
3M/(1/4")	2048	1536	1.75	60	2.8	2.20	1.59
3M/(1/3.5")	2048	1536	2	60	2.8	2.51	1.81
3M/(1/3.2")	2048	1536	2.2	60	2.8	2.76	2.00
5M/(1/3.2")	2592	1944	1.75	60	2.8	3.52	2.01
5M/(1/2.5")	2592	1944	2.2	60	2.8	4.42	2.53
8M/(1/2.5")	3264	2448	1.75	60	2.8	5.58	2.53

Note: Underlined values indicate which numbers to be used as HFDs.

level of the MTF, and by the edge and color artifacts introduced during the FR procedure. To consider the resolvable power in object space for cameras with an ideal diffraction-limited lens, we assume that the SNR of the sensor is high enough and the edge and color artifacts are small enough not to affect the resolvable power of the processed image. We also use the following criteria for defining the reliable detection of the smallest feature of a regular-printed pattern or text: (1) the minimally resolvable feature of the image should be at least the size of a pixel and (2) the geometrical blur of a point source image should be contained within a pixel element. That condition corresponds to the object distance range:

$$\left| \frac{1}{x_1} - \frac{1}{x_0} \right| < \frac{1}{\text{HFD}} \tag{4.10}$$

The minimally resolvable feature in object space (Δ) would be restricted by the size of the projection of the effective pixel size into the object space:

$$\Delta = a_e \left(\frac{x_1}{F} \right), \quad \text{when} \quad \left| \frac{1}{x_1} - \frac{1}{x_0} \right| < \frac{1}{\text{HFD}}$$

$$\Delta = \frac{F}{f\#} \times \left| \frac{1 - x_1}{x_0} \right|, \text{otherwise} \tag{4.11}$$

When expressed as a fraction of the size of a camera aperture D, the normalized resolvable object feature size Δn will be

$$\Delta n = \frac{x_1}{\text{HFD}} \quad \text{when} \quad \left| \frac{1}{x_1} - \frac{1}{x_0} \right| < \frac{1}{\text{HFD}}$$

$$\Delta n = \left| \frac{1 - x_1}{x_0} \right|, \text{otherwise} \tag{4.12}$$

The minimal value of Δn is achieved at $x_1 = x_{1\text{min}}$, and is equal to

$$\Delta n_{\text{min}} = \frac{x_0}{\left(\text{HFD} + x_0 \right)} \tag{4.13}$$

As an illustrative example, the three curves in Figure 4.9 were drawn for a 3 MP module with the lens at three best focus

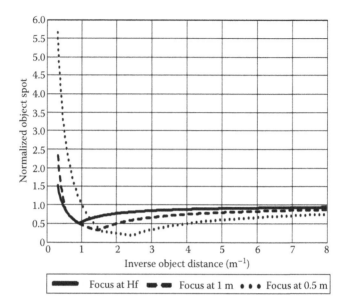

FIGURE 4.9 Resolvable spot in object space normalized to the lens aperture. Computations for the 1.75 μm 3 MP module: *Solid line*, Camera best focus is at HFD (2.195 m). The minimal object spot is reached at distance HFD/2 and is equal to *D*/2. Faraway objects stay in focus. *Dashed line*, Best focus distance is at 1 m. *Dotted line*, Best focus distance is at 0.5 m. Note how faraway objects get out of focus at longer distances.

distances: the *solid curve* corresponds to the distance of 1 HFD; the *dashed curve* corresponds to 0.455 HFD; and the *dotted curve* corresponds to 0.228 HFD. We know that when a standard imaging module is used for general-purpose applications in outdoor-indoor photography, the optics is typically adjusted so that the best-focus distance is at HFD. This ensures that the images of faraway object will remain sharp, while providing optimal coverage of the scene's DOF range.

When the lens is focused at HFD defined by the effective pixel size, the smallest resolvable feature size in object space, according to Equation 4.13, is always larger or equal to 1/2 of the lens aperture. Comparing the behavior of the three curves in Figure 4.9, we can see that the dotted curve obviously gives better object resolution than the solid curve, but that improvement comes at the price of a significant decrease in the resolution at faraway distances. Because of the soft image at infinity, the overall performance of the fixed-focus camera with this best focus setting will be judged poorly.

The restoration of the image sharpness by applying the methods of EDOF technology can be interpreted, from a system point of view, as an effective reduction in the HFD of the imaging module. When the range of the in-focus object distances x_{1min} to x_{1max}, as defined by Equation 4.7, is extended by a factor M, the best-focus object distance, x_0, could be shifted to the position HFD/M, and the objects at infinity would still remain in-focus after the FR processing. In effect, it could be simply interpreted as if the HFD of the camera was reduced by a factor M, with all other parameters remaining the same.

The graphs in Figure 4.10 illustrate the change in the size of a minimally resolvable object feature for the 1.75 μm 3 MP module with an EDOF extension factor of $M = 2$. The solid curve corresponds to the resolvable spot size when the best-focus position stays at HFD. There are improvements in comparison with the standard camera, but the full extension range is not realized. To take full advantage of the EDOF extension factor, the lens should be adjusted for the best-focus distance at HFD/2. That would allow achieving the minimal resolvable spot equal to 1/4 of the lens aperture at the distance HFD/4, as illustrated by the dashed curve.

4.2.4 EDOF Extension Factor Requirements for Text Reading

In practice, the size of the minimally resolvable object feature is usually defined by the specific camera application. For example,

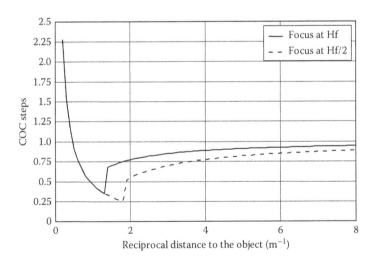

FIGURE 4.10 Resolvable spot in object space normalized to the lens aperture for the 1.75 μm 3 MP module with an EDOF extension factor $M = 2$: *Solid line*, Best focus distance is at HFD (2.2 m). *Dashed line*, Best focus distance is at HFD/2 (1.1 m). The minimal object spot is reached at distance HFD/4 and is equal to D/4. Faraway objects stay in focus. Note: the sharp step in the graph is caused by insufficient accuracy of the EDOF spot blur model.

the typical requirement for resolving printed text material is to resolve the feature size of 0.4 mm. This means that the EDOF extension factor M of any given imaging module used in an OCR application should be sufficient to produce the needed resolvable spot size. As a result of this consideration, the required EDOF factors for various types of imaging sensors, used in mobile handset applications, are presented in Table 4.2. It is assumed that the optical system of the module provides a full FOV of 60°, has a lens stop $F2.8$, and the required minimal resolvable feature in the object space is 0.4 mm.

It is interesting to note that the low-resolution modules have the natural advantage of resolving smaller object features due to very short HFDs. At the same time, the high-resolution module at the 1 MP count and above may need significant focus extension factors. The data in Table 4.2 were computed for the case when imaging optics provides a good enough resolution to take full advantage of the sensor pixel resolution. In practice, we often observe some tradeoff in the image quality, so that the effective resolvable pixel count of the image is significantly smaller than that of the sensor.

4.3 Application of SQF for EDOF Imaging

In pictorial imaging, the DOF parameter is defined not purely by a single pixel resolution, but more by an observer perceptual phenomenon with visual normalization to the sharpest part of the image. To provide a metric of perceived sharpness in pictorial images, a set of psychophysical tests was performed on a wide range of MTF shapes (Granger and Cupery 1972). The criteria developed as a result of the tests, named the subjective quality factor (SQF), have been found useful in determining the DOF extension factors of MP imaging systems, and identifying the EDOF lens efficiency, the FR efficacy, as well as the best focus positions of EDOF lenses. SQF is defined by the equation:

$$\text{SQF} = 100 \cdot \frac{\int_{v-}^{v+} M_s(v)\,d(\ln v)}{\int_{v-}^{v+} d(\ln v)} \tag{4.14}$$

TABLE 4.2 Required EDOF Gains and Minimal In-Focus Distances for Different Cameras

Sensor Type/ (Format)	Pixel (µm)	HFD (m)	F (mm)	D (mm)	Resolvable Feature (mm)	EDOF Gain Needed	Minimum In-Focus Distance (m)
VGA/(1/11")	2.2	0.27	1.53	0.54	0.27	1.00	0.13
VGA/(1/6")	3.6	0.44	2.50	0.89	0.45	1.11	0.20
VGA/(1/4")	5.6	0.69	3.88	1.39	0.69	1.73	0.20
1.3M/(1/6")	1.75	0.86	2.43	0.87	0.43	1.08	0.40
1.3M/(1/5")	2.2	1.08	3.05	1.09	0.54	1.36	0.40
1.3M/(1/4")	2.8	1.37	3.88	1.39	0.69	1.73	0.40
1.3M/(1/3")	3.6	1.77	4.99	1.78	0.89	2.23	0.40
2M/(1/5")	1.75	1.34	3.03	1.08	0.54	1.35	0.50
2M/(1/4")	2.2	1.69	3.81	1.36	0.68	1.70	0.50
2M/(1/3.2")	2.8	2.15	4.85	1.73	0.87	2.17	0.50
3M/(1/4")	1.75	2.20	3.88	1.39	0.69	1.73	0.63
3M/(1/3.5")	2	2.51	4.44	1.58	0.79	1.98	0.63
3M/(1/3.2")	2.2	2.76	4.88	1.74	0.87	2.18	0.63
5M/(1/3.2")	1.75	3.52	4.91	1.75	0.88	2.19	0.80
5M/(1/2.5")	2.2	4.42	6.18	2.21	1.10	2.76	0.80
8M/(1/2.5")	1.75	5.58	6.19	2.21	1.10	2.76	1.01

Note: Resolvable object feature is 0.4 mm.

where $M_s(v)$ is the full imaging system MTF, $v-$ and $v+$ correspond to the 3 and 12 c/deg frequency range as measured at the retina. At the viewing distance of two picture heights (a rather challenging condition for screen observation), $v-$ and $v+$ will map to 41–163 c/mm in the sensor plane for a 2 MP 1.75 μm sensor. The SQF is the mean modulation transfer over this range with logarithmic frequency weighting. Log v weighting is sometimes interpreted as the amplitude of the average pictorial scene noise power spectrum (NPS). The perceptual sharpness of the image can be expressed in units of SQF at various object distances.

The interesting property of the SQF parameter in relation to diffraction-limited lens resolution is revealed when expressing it as the function of inverse object distance normalized by the HFD of the lens (defined by a COC of 1/1300 of image size diagonal). The SQF becomes independent of the focus and the aperture stop (Bakin and Keelan 2008). The SQF dependence from normalized object distance, for different diffraction-limited lenses will fit to the same curve. Figure 4.11 shows SQF curves for a standard lens along with a lens exhibiting a strong EDOF extension factor. Moreover, by applying the Relay resolution criteria to the image

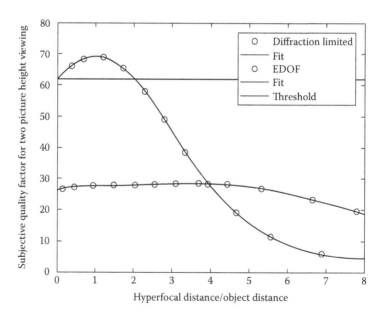

FIGURE 4.11 Through-focus SQF curves as a function of inverse object distance normalized by the hyperfocal distance of the lens for the diffraction-limited lens (*upper curve*) and the lens with the EDOF extension factor 4 (*lower curve*). The horizontal line defines the DOF threshold for the diffraction-limited lens.

created with the diffraction-limited lens, we can identify the acceptable range of SQF drop when objects are at HFD. It will also correspond to image sharpness reduction as expressed in units of SQF. It was found that the DOF range could be characterized by the drop of 7 SQF units. As such, it provides natural cutoff criteria for identifying the in-focus range for a lens. The cutoff criterion is shown as the horizontal line in Figure 4.11. The SQF curves for different lenses with various EDOF properties can be compared against the performance of the diffraction-limited lens. The EDOF factor of the lens could be identified by the range of distances at which the SQF drops by a value of 7 units. The lower curve in Figure 4.11 corresponds to the specially designed lens with an EDOF extension factor of approximately 4.

4.4 Image Recovery Process

4.4.1 Convolution Operation

The complete recovery of the original image created with EDOF optical elements involves the application of some sort of inverse digital filter to the intermediate image. Moreover, the EDOF element is typically designed to optimize the digital image restoration process. In such an integrated computational imaging system, the optical transfer function (OTF) or the PSF is purposely blurred to become more invariant with the object distance. The blurred images are then digitally recovered using the invariant PSF. There is no light loss due to decreased aperture. Mathematically, the PSF is the response of an optical system to a pulse input (Figure 4.12).

Consequently, image "blurring" is the result of a convolution between the PSF (x,y) and the "original" object image $O(x,y)$. The process of recovering the undistorted (single) image of the object from the output image blurred by the system OTF is based on the reverse conversion of the convolution integral. The methodology of the process is quite forward, and is based on the mathematical

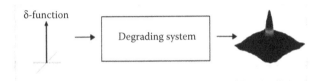

δ-function

Degrading system

FIGURE 4.12 Illustration of the PSF as the response of an optical system to impulse input.

convolution theorem (Goodman 2005). First, by definition of image formation, an output image $I(x,y)$ of the optical system is expressed as a convolution of a single image $O(x,y)$ with a linear invariant PSF $P(x,y)$:

$$I(x,y) = \int\int_{-\infty}^{\infty} O(\xi,\eta)P(x-\xi,y-\eta)d\xi d\eta \qquad (4.15)$$

The task is to find a single image $O(x,y)$ out of a measured output $I(x,y)$ and the known system PSF $P(x,y)$. For convenience of recording, the convolution integral is designated with the special character $\otimes$. In the next step, the Fourier transform operator $F\{\}$ is applied to the function $I(x,y)$ to take advantage of the convolution theorem:

$$F\{I(x,y)\} = F\{P(x,y)\otimes O(x,y)\} = F\{P(x,y)\}F\{O(x,y)\}$$
$$(4.16)$$

which states that the Fourier transform of a convolution of functions $P(x,y)$ and $O(x,y)$ is simply a product of the Fourier transforms of these functions. This procedure allows converting the integral operation into the product operation. The Fourier transform of the output image provides the angular frequency spectrum of the image. The Fourier transform of a PSF, which is by definition, the OTF of the optical system $S(f_x,f_y)$:

$$F\{P(x,y)\} = S(f_x,f_y) \qquad (4.17)$$

Now it seems obvious that the angular spectrum of the original image can be obtained by dividing the system image spectrum by the OTF of the imaging system. The original single image $O(x,y)$ is found from the inverse Fourier transform of its spectrum:

$$O(x,y) = F^{-1}\left[\frac{F\{I(x,y)\}}{S(f_x,f_y)}\right] \qquad (4.18)$$

The function $H(f_x,f_y) = 1/S(f_x,f_y)$ is referred to as the restoration filter, and, in this context as the "inverse filter" of the optical system. Since all functions on the right side of Equation 4.18 are known or may be computed from measured data, technically this is the solution for finding an ideal unaberrated image, and the problem looks solved.

Though the simplicity of this solution is very attractive, there are serious problems for real-world application:

1. The OTF of the optical system may take zero values at some frequencies. This may be due to diffraction effects, defocusing, aberrations, or discrete pixel sampling. When the OTF gets close to zero, the restoration filter would provide huge gains to the measured data spectrum. Since the dynamic range of the measured data is limited by the electronic readout circuits, the computed output results become unpredictable.

2. The simple inverse filter solution also does not account for the noise floor present in the system. Along with the desired signal, the transform boosts most of those frequency components that have the worst SNR. As a result, the recovered image is usually dominated by noise.

To get around the first problem, the OTF function of the EDOF system needs to be modified in such a way as to become monotonic and avoid approaching zero over the range of resolution frequencies subject to restoration and the range of object distances to be analyzed. Later, we will show how this is achieved in several examples of system designs. The solution for the second problem is based on the use of a modified restoration filter, which takes into account the noise distribution in the measured data. It is called the Weiner filter.

4.4.2 Weiner Filter

In the Weiner filter model, the measured image output has an added noise component $n(x,y)$ in the form (Goodman 2005):

$$I(x,y) = P(x,y) \otimes O(x,y) + n(x,y) \qquad (4.19)$$

Then, the distributions of average power over the frequency spectrum are assessed for the image ($\Phi_0(f_x, f_y)$) and for the random noise signal ($\Phi_n(f_x, f_y)$). The goal is to produce a linear restoration filter that minimizes the mean-square difference between the original object $O(x,y)$ and the estimate of the object through a series of iterative substitutions. The final solution for an optimized restoration filter is given by

$$H\left(f_x,f_y\right) = \frac{S^*\left(f_x,f_y\right)}{\left[\left|S\left(f_x,f_y\right)\right|^2 + \dfrac{\Phi_n\left(f_x,f_y\right)}{\Phi_o\left(f_x,f_y\right)}\right]} \qquad (4.20)$$

where $S^*(f_x,f_y)$ is a complex conjugate function of $S(f_x,f_y)$, so that

$$\left|S\left(f_x,f_y\right)\right|^2 = S^*\left(f_x,f_y\right)S\left(f_x,f_y\right) \qquad (4.21)$$

This type of filter provides a stable converging solution for the image restoration problem in digital iterative algorithms, and has become widely accepted in the industry. One can note that, in the case when the noise spectrum is negligibly small relative to the image signal ($\Phi_n/\Phi_0=1$), the optimum filter reduces to a regular inverse filter:

$$H \approx \frac{1}{S} \qquad (4.22)$$

And in the other extreme, when noise is the dominant factor in the signal ($\Phi_n/\Phi_0 \gg 1$), it simply converts to a strongly attenuating linear factor:

$$H \approx \frac{\Phi_o}{\Phi_n}S^* \qquad (4.23)$$

4.4.3 EDOF Processing Pipeline

In the system employing an EDOF processing pipeline, the raw data output from the sensor first undergo defect correction and black level subtraction operations (see Figure 4.13). The FR block is usually positioned directly after that. The computed power frequency spectrum distribution of the input image is applied to the Weiner filter. The sensor NPS is also tabulated in advance and stored in the lookup table. The FR restores the best estimation of $O(x,y)$ in a mathematical sense through digital iterative runs with an input image used as a starting point. After the convergence of the FR solution, the data are ready for further processing through the second-level steps of color processing, including demosaicing, white balancing, lens shading, color and gamma correction, contrast enhancement, sharpening, and denoising, as a starting point.

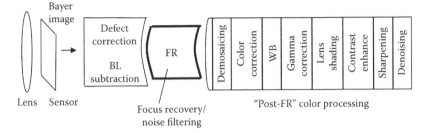

FIGURE 4.13 Schematic of an image-processing pipeline with an EDOF focus recovery block.

As the EDOF extension factor M increases, there is inevitable loss in the quality of picture. The quality of the EDOF FR is most strongly affected by:

1. PSF variations
   a. Over object distances
   b. Over field heights (image plane)
   c. Across different color channels
2. PSF-reduced peak sharpness (Strehl ratio)
3. Nonstationary noise across an image

The purpose of various EDOF methods and the FR algorithm developed in recent years is to overcome these negative factors and provide a practical robust system capable of meeting the customer demands in a very competitive mobile-camera marketplace.

4.5 Practical EDOF Methods

A number of works have addressed the development in extending the DOF in camera systems. For example, weighted zone plate apodization (Jorge and Berriel-Valdos 1990) allowed achieving a high Strehl ratio over a large defocus region with further improvement after applying digital image processing. Various depth estimation methods (Grossmann 1987; Ko et al. 2007; Bae and Durand 2007) are based on estimating high-order statistics in an image, on edge detection and blur measurement, on using multiple cameras to acquire images of a scene, or on processing multiple defocused images. Here, depth estimation based on a single image provides limited depth resolution. Also, the accuracy of a depth map could be affected by scene contents. Use of multiple cameras or multiple defocused images would require more resources. In light field photography (Ng et al. 2005), the depth extension is achieved by inserting a microlens array between the main lens and the sensor. Each microlens in the array produces an image of the exit pupil of the lens camera onto a unique area of the sensor. The camera then extracts information about both the horizontal and vertical parallaxes, which improves the reliability of the depth estimates. The drawback of this camera is its low overall resolution. There have been several attempts to achieving EDOF using wave-front coding phase masks (PMs) in combination with digital processing. Specifically, the method of cubic-phase modulation (Dowski and Cathey 1995) and apodized asphere (Chi and George 2003; Chen et al. 2008) allowed achieving large EDOF extensions. In such an integrated computational imaging system, the OTF or the PSF is purposely blurred to become more invariant with the object distance. A special purpose optical aspheric element is placed at or near the aperture stop of the imaging system. This optical element modifies the imaging system in such a way that the resulting PSF and OTF are insensitive to a range of defocus or defocus-related aberrations. The PSF and OTF are not, however, the same as that obtained with a good-quality, in-focus imaging system. By making the imaging system insensitive to defocus aberrations, the resulting images are left with a specialized, well-defined blur. This blur is removed with digital signal processing. The blurred images are then digitally recovered using the invariant PSF.

In another EDOF method of multifocus chromatic aberration (Tisse et al. 2008), the lens design intentionally provides an enhanced longitudinal chromatic aberration, so that each color plane has its focus at a different shifted object distance and thus extends the focusing range. The last three of the described technologies are covered in more details below.

4.5.1 Wave-Front Coding

4.5.1.1 Cubic-Phase Modulation

Wave-front coding with cubic PMs works to blur the image uniformly, using a waveplate with a cubic profile. Digital image processing then removes the blur and introduces noise depending upon the physical characteristics of the processor. The dynamic range is sacrificed to extend the DOF depending upon the type of digital restoration filter used. The mask can also correct optical aberration (Dowski and Cathey 1995).

This technique was pioneered by Edward Dowski and Thomas Cathey at the University of Colorado in the United States in the 1990s (Dowski and Cathey 1995). After the university showed little interest in the research (http://www.tutorgigpedia.com/ed/Wavefront_coding-_note-KeepingFocus), Dowski and Cathey founded a company to commercialize the method, called CDM-Optics. The company was acquired in 2005 by OmniVisionTechnologies, which has released wave-front coding-based mobile-camera chips as TrueFocus sensors.

A block diagram of wave-front coding imaging systems is shown in Figure 4.14. The optical section is a traditional optical system modified with a generalized aspheric wave-front coding optical element placed near the aperture stop. The addition of this optical element in the imaging system results in images with a specialized well-defined blur or PSF that is insensitive to defocus. Digital processing applied to the sampled image produces a sharp and clear image that is very insensitive to defocus effects.

A ray-based explanation of wave-front coding is shown in Figures 4.15a and 4.15b. Figure 4.15a shows rays from an ideal traditional lens focusing parallel light. These rays are converging to the focal point at the optical axis. Figure 4.15b shows the rays from the imaging system after it has been modified with a simple cubic-phase surface. Light rays from the modified lens no longer travel toward a point of best focus, but travel so that the distribution of the rays is very insensitive to the position of the image plane. The image of a point of light will not be a point image with the modified or wave-front coded system, but will be a specialized blur. Digital signal processing on the image is required to remove this blur.

A ray-based explanation of wave-front coding is shown in Figures 4.15a and 4.15b and is reproduced here from the work

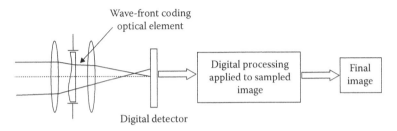

FIGURE 4.14 Block diagram of wave-front coding imaging systems. The optical section is a traditional optical system modified with a generalized aspheric wave-front coding optical element placed near the aperture stop. (From Dowski, E. R. and Johnson, G. E., *Proceedings of SPIE 3779*, 137–145, 1999. With permission.)

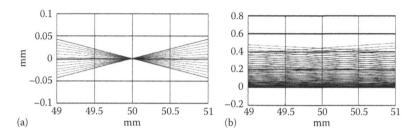

FIGURE 4.15 Ray-based explanation of wave-front coding. An ideal traditional focusing lens produces a ray density similar to (a) near the best-focus image plane and (b) ray distribution near the focal plane after modifying the traditional imaging system with a simple cubic-phase mask. (From Dowski, E. R. and Johnson, G. E., *Proceedings of SPIE 3779*, 137–145, 1999. With permission.)

of Dowski and Johnson (1999). An ideal traditional lens focusing parallel light produces a ray density similar to (a) near the best-focus image plane. After modifying the traditional imaging system with a simple cubic-phase surface, the ray density near the image plane will look like (b). The ray density of the traditional system is maximally sensitive to movement of the image plane or defocus. The ray density of the wave-front coded system, in contrast, is seen to be very insensitive to the location of the image plane. The ray density of the wave-front coded imaging system also shows that nowhere a sharp image is formed behind the lens. Digital processing is required to restore the sharpness based on the known blur of the PSF.

In mathematical formulation, the problem is stated as finding the PM profile that minimizes the variation of the OTF due to image defocusing over a predetermined range of object distances. The resulting PSF is not itself a point. But, as long as the OTF does not contain zeroes, image post-processing may be used to correct the PSF and the OTF such that the resulting PSF is nearly identical to the in-focus response of a standard optical system.

In wave optics methodology, a wave converging to the point L_2 behind the lens may be interpreted as a quadratic-phase distribution across an exit pupil aperture:

$$\varphi(x,y) = -\frac{\pi}{\lambda L_2}\left(x^2 + y^2\right) \qquad (4.24)$$

The defocus phase-error function $W(x,y)$ is the difference between the phase distribution of an ideal wave converging at focus L_2 and at some other point L_i:

$$W(x,y) = -\frac{\pi}{\lambda L_2}(x^2 + y^2) + \frac{\pi}{\lambda L_i}(x^2 + y^2) \qquad (4.25)$$

The maximum value of the error is accumulated toward the outer diameter (D) of the aperture:

$$W_m = -\frac{\pi}{\lambda}\left(\frac{1}{L_2} - \frac{1}{L_i}\right)\frac{D^2}{4} \qquad (4.26)$$

In the work of Dowski and Cathey (1995), it was proposed that the modification of a standard optical system by a rectangular phase-modulation mask of cubic profile would allow to obtain an analytical solution for OTF insensitive to defocus errors over a predetermined range of distances. Conceptually, simple filtering techniques applied to the intermediate images obtained with the mask would restore images with a high resolution and a large DOF. The cubic-phase modulation mask profile, in normalized coordinates, is given by

$$\theta(x) = \alpha(x^3 + y^3) \qquad (4.27)$$

where constant α controls the phase deviation ($\alpha \gg 20$). In the presence of the mask, the OTF of the incoherent system takes the form:

$$S(u, W_m) \approx \left(\frac{\pi}{12\alpha u}\right)^{\frac{1}{2}} \exp\left(j\frac{\alpha u^3}{4}\right) \exp\left(-j\frac{W_m^2 u}{3\alpha}\right) \qquad (4.28)$$

where $j = \sqrt{-1}$, and u is a normalized spatial frequency expressed in units of

$$u_0 = \frac{D}{2\lambda L_i} \qquad (4.29)$$

Because of the diffraction effects in the lens aperture, the maximum value of $u \leq 2$.

We can see that the amplitude of the OFT is independent of the defocus error W_m, while the second term of the complex phase shows dependence from defocus. It is minimized when α is sufficiently large. The computed dependences of the OTF magnitude as a function of the normalized frequency for three values of defocus error parameters 0, 15, and 30 are shown in Figures 4.16a and 4.16b and are reproduced from the work of Dowski and Cathey (1995). Figure 4.16a shows the computed dependences for the system with a cubic PM, and Figure 4.16b shows the computed dependences for a standard optical system. They clearly show that the OTF stays nearly constant for the system with a cubic PM, while it changes dramatically for the standard system. It is this stability of the OTF over the range of defocusing errors that allows the use of one focus-independent digital filter to restore intermediate images. Figures 4.17a through 4.17d show measured PSFs with traditional (a, b) and wave-front coded (c, d) imaging systems. The in-focus PSF of the traditional imaging system (a) and the out-of-focus PSF (b) differ greatly and result in a loss of contrast. In comparison, the corresponding in-focus (c) and out-of-focus (b) PSFs of the wave-front coding imaging systems are nearly identical to each other. This makes the wave-front coding imaging system very insensitive to a range of defocus.

The two-dimensional PSFs of a wave-front coded imaging system are shown in Figures 4.17c and 4.17d. These PSFs are from a rectangularity separable wave-front coded system. Notice that the circular PSFs from the traditional imaging system in Figures 4.17a and 4.17b change drastically with defocus. On the other hand, the PSFs from the wave-front coded imaging system show almost no noticeable change. Digital processing to remove the

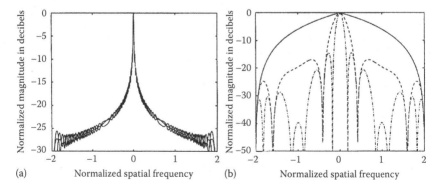

(a) Normalized spatial frequency   (b) Normalized spatial frequency

FIGURE 4.16 Magnitude of the OTF function, in decibels, versus the normalized spatial frequency in the presence of defocus errors W_m of 0 (*solid curve*), 15 (*dashed curve*), and 30 (*dashed-dotted curve*): (a) system with a cubic-phase mask $\alpha = 90$ (vertical scale is 30 dB); (b) a standard optical system (vertical scale is 50 dB). (From Dowski, E. R. and Cathey, W. T., *Applied Optics* 34, 1859–1866, 1995. With permission.)

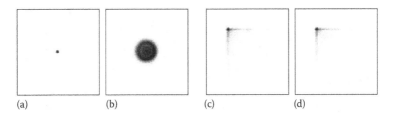

FIGURE 4.17 Measured PSFs with traditional (a, b) and wave-front coded (c, d) imaging systems for in-focus (a), (c) and out-of-focus (b), (d) objects. (From Dowski, E. R. and Johnson, G. E., *Proceedings of SPIE 3779*, 137–145, 1999. With permission.)

blur applied to a defocused traditional imaging system would require the processing to be dependent on the amount of defocus present in different areas of the image.

This is further illustrated in Figures 4.18a through 4.18f for images of a spoke target obtained with the cubic PM optical–digital system, along with images from the standard optical system (Dowski and Cathey 1995). The cubic PM optical–digital system includes both the formation of the incoherent intermediate image and the focus-independent digital filtering of this image. Without digital filtering, the intermediate images would be unrecognizable. The digital filter used for this example was a simple inverse filter that was derived from the intermediate OTF of approximation (Equation 4.28).

The left column of this figure simulates imaging of a spoke target with a standard optical system under varying degrees of defocus. The right column shows a simulation of the same imaging conditions with the cubic PM optical–digital system. The image of the spoke target from the standard system is severely degraded for even mild defocus. The images from the cubic PM system are essentially constant, and the image quality is nearly the same as that from the standard system with no defocus. Only a single digital filter is used throughout all defocus image restoration for the cubic PM system. At the same time, no single filter can be applied to the images from the standard system to correct for the effects of defocus. These simulations assumed a noise-free optical–digital system. In practice, the restoration of the intermediate image through digital filtering will alter the noise properties of the final image. As in other restorative schemes, a SNR or dynamic-range premium is required at the image. Different filtering schemes require different SNR premiums. The simple inverse filtering used here requires the

(a) (b)

(c) (d)

(e) (f)

FIGURE 4.18 Simulated images of a spoke target from a standard optical system (first column) and a cubic PM optical–digital system (second column). (a), (b) Geometrically in focus; (c), (d) mild defocus; (e), (f) extreme defocus. (From Dowski, E. R. and Cathey, W. T., *Applied Optics* 34, 1859–1866, 1995. With permission.)

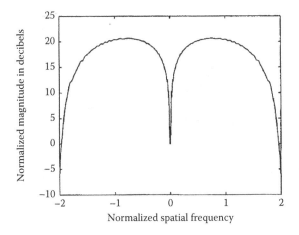

FIGURE 4.19 Magnitude of the digital filter transfer function used in simulations of the cubic PM optical–digital system. (From Dowski, E. R. and Cathey, W. T., *Applied Optics* 34, 1859–1866, 1995. With permission.)

largest premium. It has a transfer function, which is shown in Figure 4.19 (Dowski and Cathey 1995). The zero spatial frequency component of this filter is normalized to unity. With this filter the maximum magnification of any spatial frequency component is approximately 20 dB. An exaggerated estimate of the required SNR premium for this simple filter is then approximately 20 dB.

4.5.1.2 Example Applications

Several examples of an application wave-front coding system in machine vision and biometrics are shown in Figure 4.20. Figure 4.20a through 4.20f shows six images comparing the performance of a traditional imaging system (a–c) to a system using wave-front coding (d–f), for a generic binary input image (Dowski and Johnson 1999). The horizontal, vertical, and diagonal lines simply represent various constructs as may be seen in labels, bar codes, or machined parts. The center images for both systems (b, e) show the in-focus or best-focus image. The images to the left of center (a, d) give images for an object that is 35 mm from the best-focus position, having been moved away from the camera. The images to the right of center (c, f) show the results for the object at 35 mm from the best-focus position, this time having been moved toward the camera. The wave-front coded images (d–f) clearly show an increase in the DOF over the traditional system, providing an overall nominal DOF of >±35 mm.

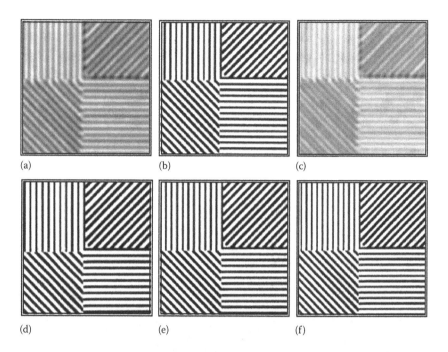

(a)   (b)   (c)

(d)   (e)   (f)

FIGURE 4.20 Example images from a machine vision/label reading application for a generic binary input. The horizontal, vertical, and diagonal lines simply represent various constructs as may be seen in labels, bar codes, and machined parts. Images from a traditional imaging system are shown in (a), (b), and (c). Images from a wave-front coded system are shown in (d), (e), and (f). The best focus images are (b) and (e). Movement of the target 35 mm away from best focus, away from the lens in image (a) and toward the lens in (c), results in badly defocused images. Movement of the target with the wave-front coded system, (d) and (f), results in little noticeable change in the images. (From Dowski, E. R. and Johnson, G. E., *Proceedings of SPIE 3779*, 137–145, 1999. With permission.)

Figure 4.21 shows six images comparing the performance of a traditional imaging system (a–c) to a system using wave-front coding (d–f) for a biometrics application. Here, the image of a fingerprint may be used in a biometric identification scheme for security. Again, the center images for both systems (b, e) show the best-focus image. The images to the left of center (a, d) give images for the finger which has been translated 40 mm from the best-focus position (away from the camera) and the images to the right of center (c, f) show the results for the finger translated 40 mm from the best-focus position (toward the camera). Again, the wave-front coded images (d–f) clearly show an increase in performance over the traditional system, providing a DOF increase of 4× to 5× over the traditional system. While with Figures 4.21a or 4.21c, one may argue that the traditional imaging system still

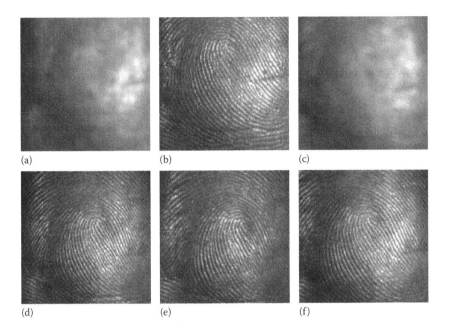

FIGURE 4.21 Biometric imaging of a fingerprint. Images with a traditional imaging are shown in (a), (b), and (c). Images with a wave-front coded imaging system are shown in (d), (e), and (f). The best focus images are shown in (b) and (e). Movement of the finger 40 mm from best focus results in images (a) and (d) (away from the lens) and (c) and (f) (towards the lens). The out-of-focus traditional images have very little spatial resolution. (From Dowski, E. R. and Johnson, G. E., *Proceedings of SPIE 3779*, 137–145, 1999. With permission.)

provides some resemblance to the object (albeit quite poor), in the biometric case the traditional imaging system barely produces recognizable images.

Wave-front coded imaging systems can have significant performance advantages over traditional imaging systems in special applications. The most obvious advantage is the greatly increased DOF/depth of focus when a traditional optical system is modified with a wave-front coding optical element and signal processing. Some enhancements enabled by wave-front coding are:

■ An inexpensive plastic lens incorporating wave-front coding into the design can produce an image quality equivalent to optics of much greater complexity and expense, since the digital processing step not only removes defocus, but also minimizes the effects of chromatic aberration, Petzval curvature, astigmatism, spherical aberration, and temperature-related defocus.

- As the MTF improvement practical limit is set by the system SNR, improvement factors of 2 in the higher spatial frequencies are usually possible.
- Imaging systems with wave-front coding can be highly customized for special applications at much less cost than a custom-designed lens system. For some applications, wave-front coding might be the only way to meet all of the requirements.

The range of EDOF extension factors in systems with a wave-front coding plate is limited only by the SNR characteristics of the sensor. One of the drawbacks in the practical implementation of a cubic plate design is the use of nonrotationally symmetrical optical elements. Though the rectangular separation of cubic profiles in the x- and y-axis allows an elegant analytical solution for the PSF function, it also creates problems in the manufacturing, alignment, and testing of the system in production. To overcome this drawback an alternative solution for the EDOF system design was developed utilizing rotationally symmetrical optical elements.

4.5.2 Aspheric Phase Apodization Method

Another practical system solution to EDOF design was implemented with the use of a rotationally symmetrical optical element having an aspheric profile with a control amount of spherical aberration built in. Again, the target is to achieve PSF invariance over a predetermined range of object distances. In this case, the initial PSF invariance is achieved on-axis with the use of a logarithmic asphere method (Chi and George 2003). The PSF is purposely blurred so that it becomes invariant with the object distance. The lens with a logarithmic asphere profile has circular symmetry with a controlled continuous radial variation of the focal length. Each radial ring has one specific local focal length which provides a sharp focus of a specific object distance on the detector plane. The rings are selected in such a way as to contribute uniformly to the integral PSF over the preselected range of distances. Therefore, the images for objects within the EDOF range are always sharply focused by one specific ring of the lens and defocused by the other rings. This method introduces a control amount of spherical aberrations in the range of up to 7λ. The obtained phase-profile function is applied to the surface near the system stop position, which allows minimizing additional third-order off-axis aberrations such as coma, astigmatism, and Petzval curvature.

In subsequent steps, an iterative optical ray-tracing algorithm is applied to achieve the PSF depth invariance for off-axis positions (Chen et al. 2008). The final PSF is typically a slow variable function of the image coordinates. The PSF data are used for generating a Weiner filter for the FR algorithm.

4.5.2.1 Lens Design Algorithm

The initial aspheric profile of the lens is obtained from applying the concept of an equal optical path (Fermat principle) to the rays following through different radial segments of the lens in the presence of the phase-variation profile. Following the original work of Chi and George (2003), the lens is divided into annular N rings of different focal lengths with equal area. The conditions of a radius ρ_n for the nth ring are as follows:

$$\rho_n = \left(\frac{n}{N} \right)^{1/2} R \qquad (4.30)$$

where R is the outer radius of the lens, as indicated in Figure 4.22 (Chi and George 2003). Consider the imaging of a point S at $x(\rho_n)$ by the rays through the annular ring, ρ_n. To provide the condition of equal contribution through each annular segment, the interval from s_1 to s_2 is subdivided into N segments, and the $x(\rho_n)$ segment is weighted as follows:

$$x(\rho_n) = s_1 + (s_2 - s_1)n/N \qquad (4.31)$$

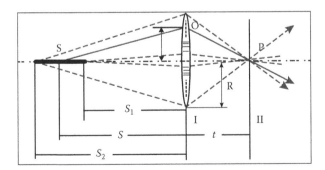

FIGURE 4.22 Lens consideration for aspheric phase optimization of system PSF. (From Chi, W. and George, N., *Journal of Optical Society of America A* 20, 2260–2273, 2003. With permission.)

Substituting ratio n/N from Equations 4.30 through 4.31 gives the basic formula linking the axial segment x with the annular ring of radius ρ:

$$x(\rho) = s_1 + (s_2 - s_1)\rho^2 / R^2 \qquad (4.32)$$

In the presence of a phase delay $\varphi(\rho)$, the total optical length L through each radial segment is given by

$$L = \sqrt{\rho^2 + x^2} + \frac{\varphi(\rho)\lambda}{2\pi} + \sqrt{\rho^2 + t^2} \qquad (4.33)$$

where t is the distance from lens plane I to plane II and λ is the free-space wavelength.

The requirement of equal optical pass lengths through each annular segment gives the condition $\partial L/\partial\rho = 0$, which provides the integral equation for the phase-delay function:

$$\Delta\varphi(\rho) = -\frac{2\pi}{\lambda}\int_0^\rho \frac{r}{\sqrt{r^2 + t^2}} + \frac{r}{\sqrt{r^2 + \left[s_1 + (s_2 - s_1)r/R^2\right]^2}} dr \quad (4.34)$$

The function $\Delta\varphi(\rho)$ could be integrated directly to provide a logarithmic component in the phase-delay distribution for the transmission function of a lens which focuses an extended portion of the optical axis into a single image point.

$$\Delta\varphi(\rho) = -\left\{ \frac{2\pi}{\lambda_0}\left(\sqrt{\rho^2 + t^2} - t\right) + \frac{\pi}{\lambda_0}\frac{R^2}{s_2 - s_1} \right.$$
$$\times \left[\ln\left\{ 2\frac{s_2 - s_1}{R^2}\left[\sqrt{\rho^2 + \left(s_1 + \frac{s_2 - s_1}{R^2}\rho^2\right)^2}\right.\right.\right.$$
$$\left.\left.\left. + \left(s_1 + \frac{s_2 - s_1}{R^2}\rho^2\right)\right] + 1 \right\} - \ln\left(4\frac{s_2 - s_1}{R^2}s_1 + 1\right) \right] \right\}$$

$$(4.35)$$

The first term gives the phase distribution for an ideal lens with a focus at distance t; the second term introduces the contribution from the controlled aspheric aberration. If the phase-delay expression is further expanded as a power series of ρ^2 and expressed in the general terms of coefficients depending on a preselected EDOF range and the lens parameters s_1, s_2, R, and t, it turns out that the main terms are ρ^4 and ρ^6, which represent the third-order and fifth-order spherical aberrations (Chen et al. 2008).

With the first-order optics parameters and a predetermined EDOF range for object distance and Equation 4.1, it becomes possible to calculate the phase-delay function of a logarithmic asphere phase plate and convert it into a surface profile. The optical surface closest to the stop is modified, as discussed earlier.

Minor changes of the aspheric coefficients for ρ^4, ρ^6, and ρ^8 do not affect the first-order optics parameters of the initial standard lens. As was the case with the cubic-phase plate design, the limit on the extension coefficient M is imposed by the dynamic range of the sensor and the acceptable level of SNR. To understand the improvements provided by the apodized aspheric profile, let us look at the PSF variation for a standard lens with a focal length of $F = 3.75$ mm and an aperture stop F2.8. The actual layout of the lens is shown in Figure 4.23.

The PSF profiles of a standard lens are shown in Figures 4.24a through 4.24c for object distances (a) 10, (b) 0.65, and (c) 0.3 m, When an object is at 10 m, the PSF has almost a diffraction-limited behavior with FWHM = 1.65 μm and a Strehl ratio of 0.879; at a distance of 0.65 m, the PSF expands to FWHM = 4.5 μm and the Strehl ratio drops to 0.165, while at an object distance

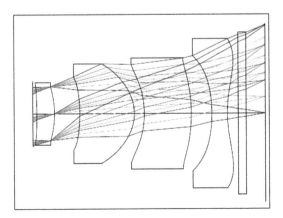

FIGURE 4.23 The layout of the standard imaging lens $F = 3.75$ mm, aperture stop F 2.8, and field of view 61°.

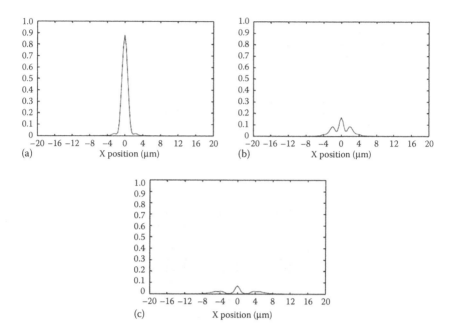

FIGURE 4.24 On-axis PSF behavior for a standard lens: (a) object distance 10 m, FWHM = 1.65 μm, and Strehl ratio 0.879; (b) object distance 0.65 m, FWHM = 4.5 μm, and Strehl ratio 0.165; and (c) object distance 0.3 m, FWHM = 8.5 μm, and Strehl ratio 0.071.

of 0.3 m, the PSF is stretched out to FWHM of 8.5 μm and the Strehl ratio is 0.071.

The on-axis PSF profiles for the lens with apodized aspheric aberration optimized for an EDOF operating range of 10–0.3 m are shown in Figures 4.25a through 4.25c. A comparison with standard lens PSFs shows a significant improvement in the PSF consistency for the lens with an optimized profile.

It is important to note that the parameter of the width of PSFs at half maximum cannot be used by itself to judge the effectiveness of EDOF optimization, since the side lobes of the PSF may increase in amplitude and carry a larger portion of the total power. A much more useful criterion for an EDOF improvement is the Strehl ratio of the PSF, which is defined as the ratio of the system PSF intensity to that of the diffraction-limited system having the same numerical aperture. As such, Figure 4.26 shows the dependence of the Strehl ratio from the object distance in the range 0.25–10 m for the standard system previously described.

The vertical scale of the Strehl ratio amplitude shown in Figure 4.26 is 0–1. As expected, the Strehl ratio demonstrates a dramatic

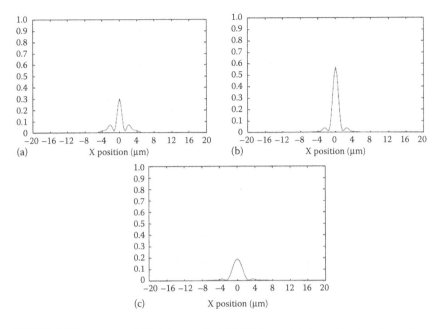

FIGURE 4.25 On-axis PSF behavior for EDOF lens with logarithmic sphere-phase apodization: (a) object distance 10 m, FWHM = 1.3 μm, and Strehl ratio 0.303; (b) object distance 0.65 m, FWHM = 1.6 μm, and Strehl ratio 0.57; (c) object distance 0.3 m, FWHM = 2.65 μm, and Strehl ratio 0.191.

drop from 0.86 to 0.06 as the object gets closer to the lens from 10 to 0.3 m, and moves out of the DOF range. For comparison, an EDOF lens with an optimized logarithmic asphere profile maintains a Strehl ratio in the range of 0.19–0.57 over the same span of object distances.

4.5.2.2 *Optimization Off-Axis Field Performance in EDOF Design*

A further improvement of the lens off-axis field performance is achieved in ray-tracing design modeling with the use of a specially designed merit function (MF). The customized MF lays emphasis on the Strehl ratio uniformity over 85% of the FOV, and works through an iterative optimization algorithm. The algorithm runs through variations in the high-order aspheric coefficients of the system's optical surfaces in order to obtain a more uniform PSF distribution over the entire FOV of the camera system, and over the range of object distances. The first-order optics parameters are constrained to be the same as before. Additional limitations are put on the allowed amount of geometrical distortion, axial color, spectrum range weights, lateral color, and maximum

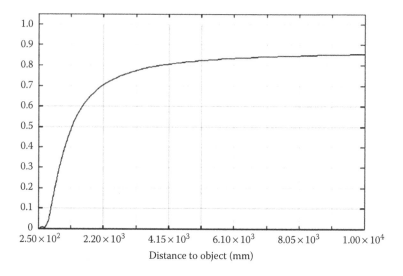

FIGURE 4.26 The axial PSF Strehl ratio versus object distance (from 250 mm to 10 m) for a standard imaging lens.

chief ray angle (CRA). For general imaging camera applications, the amount of spherical aberration, which is the combination of third-, fifth-, and seventh-order spherical aberrations is typically limited to 0.7λ based on indications from an SQF analysis, as described earlier. The improvement achieved with this "brute force" machine approach to PSF optimization for the same system as described earlier is shown in Figures 4.27a through 4.27c. The object distance range is 0.3–10 m.

The PSF profiles variations, shown in Figures 4.24, 4.25, and 4.27, demonstrate how the ray-tracing optimization algorithm allows a significantly improved PSF and Strehl ratio uniformity over a predetermined EDOF range and over the system FOV. This improvement provides a benefit to a simplified image restoration filter. It allows applying a uniform Weiner filter to the entire image for an efficient image restoration with an improved SNR. The full field image obtained for a standard camera optics with a 3 MP CMOS sensor module is shown in Figures 4.28a through 4.8c. The simulated scaled images were obtained for object distances of 0.3, 0.65, and 10 m. We can see how the resolution and contrast get lost as the object moves closer to the lens. For comparison, the images of the EDOF lens with an optimized off-axis performance are shown in Figures 4.29a through 4.29c. The same sensor and FOV are used in both sets of images.

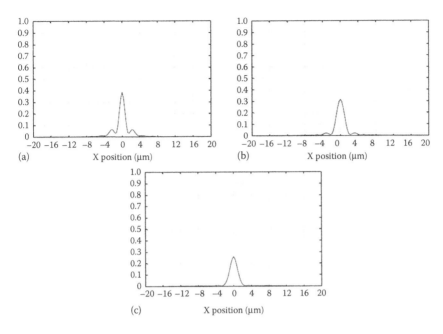

(a) X position (μm) (b) X position (μm)

(c) X position (μm)

FIGURE 4.27 PSF behavior for EDOF lens with optimized off-axis performance: (a) object distance 10 m, FWHM = 1.4 μm, and Strehl ratio 0.385; (b) object distance 0.65 m, FWHM = 2.1 μm, and Strehl ratio 0.314; and (c) object distance 0.3 m, FWHM = 2.1 μm, and Strehl ratio 0.259.

(a) (b) (c)

FIGURE 4.28 Simulated scaled images obtained with standard 3 MP camera module. Pixel size 1.75 micron, F2.8 lens, full field 61°: (a) object distance 10 m; (b) object distance 0.65 m; (c) object distance 0.3 m.

(a)   (b)   (c)

FIGURE 4.29 Simulated scaled images obtained with EDOF lens with optimized off-axis performance with a 3 MP camera module. Pixel size 1.75 μm, F2.8 lens, full field 61°: (a) object distance 10 m; (b) object distance 0.65 m; and (c) object distance 0.3 m.

When comparing the images in Figures 4.28 and 4.29, it is obvious that the image quality at 10 m is about the same for both modules. The image of the EDOF module improves slightly more than that of the standard lens module when the object is at 65 cm. The image of the EDOF module is more superior to that of the standard lens module when the object is at 30 cm. The EDOF imaging module extends the DOF to 30 cm or closer, which gives twice the extension factor of that of the standard lens module.

The achievable design performance is, as usually happens in complex optical systems, a compromise between many contributing system parameters. Some of these parameters may have contradicting demands, and the final judgment call is made based on the specific requirements of the camera application. Ideally, all the specification values should be put into the MF of a ray-tracing optimization algorithm with an assigned weighting factor to reflect the actual priorities of the design.

The Strehl ratio value of the PSF can be used as the one universal integral parameter for lens optimization. The image resolution and the overall quality are strongly affected by the size and shape of a PSF and the Strehl ratio. From the analysis of the performance of the EDOF imaging module with a 1.75 μm 3 MP CMOS sensor, it was concluded that during optimization the lens PSF and the Strehl ratio should be kept above the 0.3 threshold value. The optimized EDOF imaging modules allowed extending the FOV to 30 cm. As shown in Figures 4.28 and 4.29, the images for object distances 10 m, 65 cm and 30 cm, taken with the EDOF optimized optics, look as good as the best-focused image taken with the standard lens imaging module. An EDOF extension factor in the range of 2–3 can be achieved.

4.5.3 Multifocal Chromatic Aberration Imaging

Another successful implementation of the camera system with EDOF was demonstrated by the French company DXO (Tisse et al. 2008), using a very different approach to sharpness restoration in blurred out-of-focus images.

This is achieved with a special lens design where the longitudinal chromatic aberration has been intentionally magnified. This design approach contradicts the conventional lens design approach, in which the ideal imaging system has all aberrations suppressed. The presence of chromatic aberrations causes rays of different colors to focus at different distances from the lens. This is exactly the same effect as happens with images of out-of-focus objects, but in this case it happens in different color planes. Using the controlled increase in chromatic aberration, the image plane could be shifted so that for any distance within the EDOF, at least one color channel of a red/green/blue (RGB) image will contain the in-focus scene information (e.g., high frequencies). By determining the sharpest color (for each region in the digital image) and reflecting its sharpness on the others, it is possible to get a sharp image for all colors through the merged DOF of all three color channels. The variation of the focal length distance due to longitudinal chromatic aberrations causes near objects to appear blurred in the green and red channels, midrange objects to appear blurred in the blue and red channels, and far objects to appear blurred in the blue and green channels.

Figure 4.30a shows a typical through-focus MTF plot of a lens system with strong longitudinal chromatic aberrations. The relative positions of the best-focus planes for the RGB color channels (i.e., at the peak of each through-focus plot) represent the amount of chromatic focus shift at the center of the FOV. Figure 4.30b shows how the root mean square diameter of the blur spot within each (RGB) color channel changes as a function of the

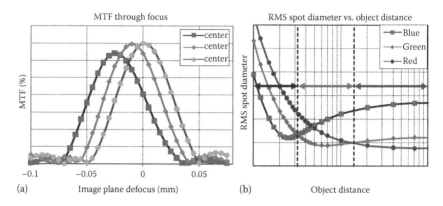

FIGURE 4.30 (a) Typical through-focus MTF plot of a lens system with strong longitudinal chromatic aberration; and (b) DOF ranges for color channels. (From Tisse, C.-L., Nguyen, H. P., Tessieres, R., Pyanet, M., and Guichard, F., *Proceedings of SPIE 7061*, 2008. With permission.)

object distance. Each color channel has its own effective DOF, as depicted by the arrows.

The color image captured through such an optical system—with at least one color channel in focus—is then processed by a sharpness transport engine. The sharpness transport engine determines, for each region of the image, the sharpest color channel and then transfers that sharpness to the other color channels, thus providing an EDOF by cumulating the effective DOF of each color channel. In the first step of generating a depth map, each pixel is assigned a depth value corresponding to a specific range of object distances. In the implementation, the scene is segmented into three coarse depth layers that are referred to as scene modes: *macro, portrait*, and *landscape*. These three modes typically coincide with the DOF of the blue, green, and red colors, respectively. This layered depth map is obtained by sorting the color channels by their sharpness level. In the second processing step, the correction of the blurred channel(s) is performed just by copying the high frequencies of the sharpest channel to the others. Unlike standard deblurring techniques, which tend to amplify noise (i.e., noise variance being amplified by the L_2 norm of the filter), sharpness transport across color channels advantageously provides a small denoising effect. This extra capability is inherent to the blending operation performed (in the mosaic domain) between pixels of different color channels. Nevertheless, noise effects may still occur depending on the illuminant. For instance, when imaging a close object under tungsten lighting, high-frequency information could be altered in the blue channel. In this case, in order to avoid noise propagation to the green and red channels, the blue channel signal is prefiltered prior to blending. After sharpness transport, the MTF will be similar in the different color channels and at the level of a traditional lens within its in-focus range. Of course, the DOF of the system will be larger (owing to the addition of the three color channels DOF). Each of the color channel MTFs is very similar to the other and does not have nulls in the spectrum.

4.5.4 Examples of EDOF Processed Images

The imaging system is able to reconstruct an MTF almost invariant (and above 50% at half Nyquist frequency) over the distance range of 0.25–4 m, and a reasonable image down to 0.15 m. This allows considering interesting applications of close imaging such as bar-code reading. Figure 4.31 shows an example of a bar-code image shot at 15 cm with the same camera module. The bar code, which has a module size of 0.25 mm, is successfully recognized by a standard bar-code reader software.

Figure 4.32 illustrates the visual benefits of an EDOF system. The same natural scene (with near and far objects) was captured with both the developed camera module and two commercial mobile-phone cameras—one is a fixed-focus camera and the other is an AF camera. The closeup images on the right of Figure 4.32

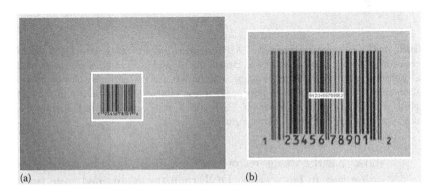

(a)   (b)

FIGURE 4.31 Example of bar-code target image resolved with DxO Digital Optics for an EDOF system. Target is located at a distance of 15 cm. The element line width is 0.25 mm. (From Tisse, C.-L., Nguyen, H. P., Tessieres, R., Pyanet, and M., Guichard, F., *Proceedings of SPIE 7061*, 2008. With permission.)

FIGURE 4.32 Comparison between 3 MP resolution pictures taken with: (a, b) a fixed-focus camera phone; (c, d) an autofocus camera phone; and (e, f) an EDOF camera module equipped with DxO Digital Optics. On the left are the entire resolution images and on the right are the cropped portions of those images (corresponding to the rectangles). (From Tisse, C.-L., Nguyen, H. P., Tessieres, R., Pyanet, M., and Guichard, F., *Proceedings of SPIE 7061*, 2008. With permission.)

show that the image restored with a multifocal chromatic aberration system has a larger DOF than those obtained from traditional imaging devices (Tisse et al. 2008). The closeup images show the EDOF offered by DxO Digital Optics. Fine details are recovered both in the foreground and in the background (Tisse et al. 2008).

4.6 What Are the Limits of EDOF FR?

In the conclusion to this chapter on EDOF technology, we will discuss the factors limiting the achievable EDOF range in imaging applications using the method of SQF discussed in Section 4.3. As was shown earlier in Figure 4.11, the diffraction-limited lens has an EDOF factor = 1 (no extension) when an SQF curve drops by 7 units from its peak. Though the FR process may eventually degrade the quality of the final image by amplifying the noise floor, it was found that to first order, the FR only shifts the SQF through-focus curves vertically without changing their shape much. So, the EDOF extension factor is determined by the initial width of flattening of the SQF curves near their peak, and is defined by the PSF blur uniformity. The amount of blur to be compensated is limited by the noise floor amplification.

The following are further findings from an analysis of the various EDOF lens designs:

- Field curvature present in the optical design takes away from the EDOF extension—typically it cannot be corrected without using significant image-processing resources.
- Axial color also takes away from the EDOF extension. Axial and lateral colors can create artifacts amplified by FR (because they resemble signal).
- Slow, smooth, monotonic change in the PSF versus the field height minimizes tile boundary artifacts.
- These factors all trade off, and the worst problem seems to dominate. With an integrated, adaptive FR and noise suppression, EDOF extension factors of 2× to 3× appear feasible.
- Ideally, an EDOF lens should blur by the maximum amount that is algorithmically recoverable without noise amplification in uniform areas, or objectionable edge artifacts.

And finally, let us see how an MTF analysis allows predicting the performance of an EDOF system in a bar-code reading application.

The MTF diagram shown in Figure 4.33 has three curves: the bottom curve indicates an MTF curve taken from the sensor before FR. It is mostly affected by the designed-in blur in the PSF; the horizontal curve simply indicates the threshold for the acceptable MTF contrast required to resolve the bar-code

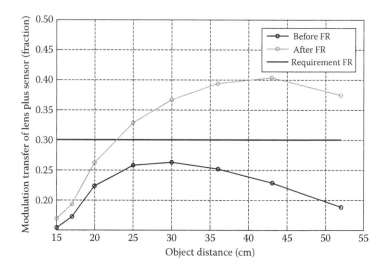

FIGURE 4.33 MTF analysis of a bar-code reading application with an EDOF camera: the center line is the threshold for bar-code readability; the lower curve is the MTF of the lens plus sensor after tolerance factors; the upper, gray curve is the MTF of the module after applying FR.

pattern of 0.5 mm element size. It is found to be at the level of 0.3 for any object distance; the upper curve is the MTF curve of the output image after the boosting spatial frequencies (along with noise) through the FR algorithm. One can see that for the particular system shown in Figure 4.31, the bar code could be reliably read only from the distances above 23 cm. If the application specifications ask for shorter reading distances, then a different design approach would be needed to make the system work.

References

Bae, S. and Durand, F. (2007), Defocus magnification, *Computer Graphics Forum* 26, 571–579.

Bakin, D. and Keelan, B. (2008), Imaging with depth extension, *Proceedings of SPIE 7061*, 706106-1-10.

Chen, X., Bakin, D., Liu, C. and George, N. (2008), Optics optimization in high-resolution imaging module with extended depth of field, *Proceedings of SPIE 7061*, 706103-1-12.

Chi, W. and George, N. (2003), Computational imaging with the logarithmic asphere: Theory, *Journal of Optical Society of America A 20*, 2260–2273.

Dowski, E. R. and Cathey, W. T. (1995), Extended depth of field through wave-front coding, *Applied Optics 34*, 1859–1866.

Dowski, E. R. and Johnson, G. E. (1999), Wavefront coding: A modern method of achieving high performance and/or low cost imaging systems, *Proceedings of SPIE 3779*, pp. 137–145.

Goodman, J. W. (2005), *Fourier Optics*, 3rd edn, Chapter 8, Roberts & Company, New York, pp. 259–265.

Granger, E. M. and Cupery, K. N. (1972), SQF—An objective metric of perceived sharpness in pictorial images, *Photographic Science Engineering 16*(3), 221–230.

Grossmann, P. (1987), Depth from focus, *Pattern Recognition Letters 5*, 63–69.

Jorge, O.-C. and Berriel-Valdos, L. R. (1990), Zone plate for arbitrarily high focal depth, *Applied Optics 29*, 994–997.

Ko, J., Kim, M., and Kim, C. (2007), 2d-to-3d stereoscopic conversion: Depth-map estimation in a 2d single-view image, *Proceedings of SPIE Applications of Digital Image Processing XXIX*.

Marr, D. and Poggio, T. (1979), A computational theory of human stereo vision, *Proceedings of Royal Society of London Series B 204*, 301–328.

Ng, R., Levoy, M., Bredif, M., Duval, G., Horowitz, M., and Hanrahan, P. (2005), Light field photography with a hand-held plenoptic camera, Stanford Technical Report CTSR 1, 11.

Tisse, C.-L., Nguyen, H. P., Tessieres, R., Pyanet, M., and Guichard, F. (2008), Extended depth-of-field using sharpness transport across colour channels, *Proceedings of SPIE 7061*, 706105-1–12.

CHAPTER 5

Liquid Lens
A Key Adaptive Component for Cameras

Bruno Berge

Contents

149

5.1 Introduction: Appearance of Variable Lenses in Cameras

Film cameras equipped with motors and automatic and motorized adjustments have existed for at least 35 years (the Konica C35 AF, known to be the first autofocus [AF] film camera, was launched in 1977) (Hartmann 1976). After many years of debating whether the best results are achieved through manual or automatic focusing, the consensus is that motorized AF is faster and more accurate in many situations, even for high-end reflex cameras. Motorized AF appeared on the market well before electronic sensors (charge-coupled device [CCD] or complementary metal-oxide-semiconductor [CMOS] sensors, see Chapter 2, this volume) replaced film cameras. When the transition from argentic to digital photography took place, few changes were made in the mechanical architecture of the camera's main optics (image-forming optics). Nevertheless, in the past decade, the transition to digital photography has triggered a trend toward miniaturization, which has drastically changed camera architectures*. Pixel sizes have shrunk considerably (the surface of the smallest pixels on the market has decreased by a factor of ≈2 every 2–3 years [Suzuki 2010]), as has the camera format: when film cameras were in use, almost any image format was above 10 mm, whereas it is common now to find camera modules with image sensors below 5 mm diagonal (the diagonal of popular 1/4-inch sensors is 4.54 mm). For such miniature cameras, the choice of manually adjusting the focus is no longer available as fingers would be too big for this. In addition, the low-end demand of the mobile market has seen the introduction of wafer-scale cameras (Wolterink and Demeyer 2010; Rudmann and Rossi 2004), an ultracompact concept, which is even more adverse to moving parts, such as AF.

Nowadays, the miniaturization leading to mobile cameras represents a size factor ranging from approximately 3 to 5 compared to compact cameras, but the technical solutions that are used in compact cameras do not work at the scale of further miniaturization because most of the guiding of optical systems in digital still cameras (DSCs) is based on the friction of sliding cylinders to move an optical subassembly with regard to other optical blocks, while keeping precise alignments. Due to the surface/volume ratio,

---

* For a modern camera module architecture, see for instance: http://www.st.com/internet/com/
SALES_AND_MARKETING_RESOURCES/MARKETING_COMMUNICATION/
FLYER/flub69400209.pdf.

further miniaturization of cameras has made this solution impractical. Thus, for at least 20 years, the industry has been trying to change the way optical adjustments are made, in order to suppress any physical motion of glass/plastic lenses. The outcome has been the incorporation of variable lenses in digital cameras (Berge and Peseux 2000).

5.1.1 Liquid Lens

What is a variable liquid lens? For centuries, scientists have combined lenses of very different optical powers, from diverging lenses to highly converging ones, to design and manufacture complex optical instruments (see Figure 5.1). Until now the optical properties of each lens, either made from glass or plastic and typically requiring months to be produced, have always been fixed. The liquid lens is thus a "smart lens" that can be reconfigured on demand with just a variation of voltage. The lens can adapt rapidly and continuously from diverging to converging and can be modeled to support AF. The liquid lens can also provide an adjustable optical tilt and support optical image stabilization (OIS) functions (see Chapter 7).

Variable lenses have been investigated for over 70 years and some basic principles are now well established, for example, a liquid chamber with deformable transparent membranes (Graham 1940) (historically the first), electroactive deformable gels (Hamada et al. 1988), liquid crystals with gradients (Commander et al. 2000; Naumov et al. 1998; Asatryan et al. 2010), and

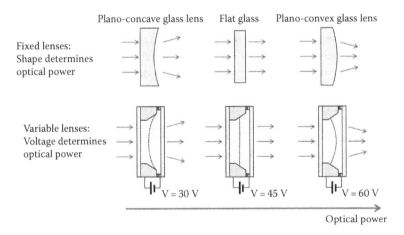

FIGURE 5.1 Variable lenses compared to fixed lenses: a voltage variation induces internal reconfiguration, leading to optical power variations.

dielectric liquid deformations (Cheng et al. 2006). All these principles were investigated but none of them led to commercial application in the field of miniature applications.* In this chapter, we will focus on liquid lenses based on electrowetting (EW), which were the first variable lenses to appear in the miniature camera industry (Berge and Peseux 2000).

The principle of the liquid lens uses two liquids having the same density: once encapsulated hermetically without air, the liquids become totally insensitive to any gravitational or translational acceleration: this makes them remarkably stable, and insensitive to shocks and vibrations. In addition, the forces to change the optical lens properties are small, as the restoring forces are essentially capillary forces, so the power consumption can be minimal (Berge and Peseux 2000).

The technology uses the principle of EW and a combination of transparent and optically defect-free liquids to create a lens and change its characteristics in real time. For 40 years, liquids have been used in optical systems for high-end products such as goggles and camcorders (Canon developed a variable prism in the 1980s, using liquids trapped between two glass windows, called the "Vari-Angle Prism Image Stabilizer": see, for instance, http://www.canon.com/bctv/faq/vari.html); however, an innovation by Varioptic (France) created a real-time programmable platform that offers to change the shape of the liquids in a very fast, repeatable, precise, and controlled way.

5.1.2 Integration of Variable Lenses in Cameras: No Moving Part

In order to make a complete camera, the liquid lens component is combined with other optical glass or plastic molded lenses. In such systems, all lenses are fixed with glue or by mechanical fastening. This realizes a no-moving-parts optical system, with the liquid lens being the adjustable part by tuning the system electronically.

Figure 5.2 shows typical examples of the integration of an AF camera with either (a) a liquid lens architecture or (b) a voice coil motor (VCM) (Wu et al. 2009), and (c) a piezo motor (Chou 2009).

5.1.3 Faster AF Strategies: Open-Loop "Point and Shoot" and Nonsequential Algorithms

A unique feature allowed by this new way of manufacturing cameras, using variable lenses instead of motors, is the ability to focus

---

* Today, lenses with deformable membranes are used industrially in ophthalmic applications, which have been developed by Joshua Silver and his team, but their specification appears to be completely different from that of miniature cameras. See, for instance, Treisman and Silver (1988).

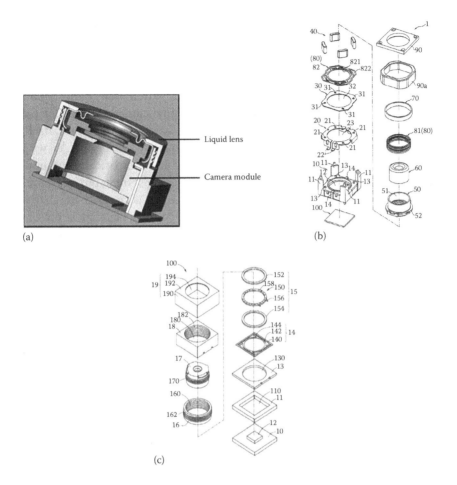

FIGURE 5.2 Comparison of the construction of a liquid lens-based camera module (no-moving-parts) (a) with VCM-based (b) and piezo-based (c) camera modules. ((a) Wu, H., Chou, C., and Shyiu, S., Voice coil driving auto-focus lens module, U.S. Patent 7612957, 2009; (b) Chou, Y., Camera module with piezoelectric actuator, U.S. Patent Application US2010/0271541 A1, 2009.)

more quickly, still keeping the required accuracy. The following two approaches have been successfully tried with variable lenses.

The first approach is to add an independent focus telemeter, which assists the camera and delivers a set focus point. This enables focusing of the camera essentially in one step, thus considerably reducing the lag of focusing. As it is independent, this system works in an open-loop mode, and thus requires very stable focusing mechanisms, which, so far, only liquid lenses based on EW have achieved in industrial products (Kawashima and Takahashi 2007). The conventional mechanical systems for focus adjustment are not accurate enough to perform an open-loop focus control.

The second approach is to use one of the specificities of variable lenses—the ability to rapidly change to arbitrary focus distances—in order to improve the AF search used in most compact cameras (see Figure 5.3). Traditional cameras search for the best focus using a sharpness score delivered by the sensor itself, through pixel-level, high spatial frequency detection. As with every closed-loop system, it is very stable, automatically compensating for all potential drifts in the focusing system, but it requires at least 15–20 focusing steps to achieve full accurate focus sharpness. Today, this is still part of the focusing lag that is present even in the most advanced mobile-phone cameras, a feature that irritates the user.

The number of steps to find the best focus is considerably decreased in variable lenses (sometimes by a factor ≈2), by using a nonsequential search inside the accessible focus range (Moine et al. 2011). An example of such a nonsequential search is the dichotomy strategy: one measures the focus score at the minimum

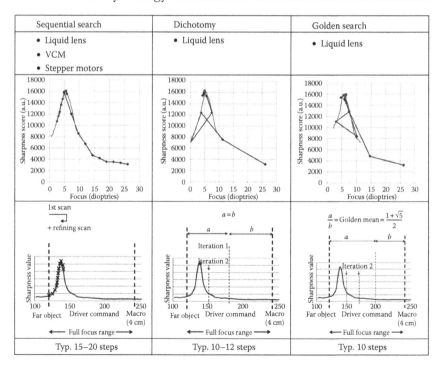

FIGURE 5.3 Improvement of the focusing search algorithm's efficiency using a camera based on a liquid lens (see Section 5.4.3).

focus, at the maximum focus, and in the middle focus, thus dividing the original search zone into two equal parts. Then, the algorithm discards the least sharp part of the original focusing range, and iterates the search on the other half.

Ultimately, it is still possible to further improve the efficiency and speed of the search strategy, mimicking strategies used in random access memory (RAM), known as the "golden search" (Press et al. 1992). The principle is to measure only two points, which are both off-centered by the golden mean inside the focus range. These two points decide which side of the focusing range is discarded for the next iteration. Although the size of the next range only decreases by a factor of 0.62 ($=\tau/(1 + \tau)$), where τ is the golden mean, this process is slightly more efficient than the dichotomy, as the extremes of the range are never measured. Nevertheless, in some situations, dichotomy algorithms can be biased to be more efficient at some object distances. For example, in consumer photography, it is well known that most pictures are taken at far object distances, and fewer are taken in the macro mode, such that the highest efficiency is achieved when far-sight is favored in the algorithm.

5.2 Principle of Operation of Liquid Lens Based on EW

5.2.1 Basic Principle

Figure 5.4 shows the basic structure of a liquid lens based on EW: a transparent cell is filled with two immiscible transparent liquids with different optical indices. The interface between the two liquids is a portion of a sphere aligned with the optical axis, thus forming a lens. The edge of the liquid–liquid interface is a line

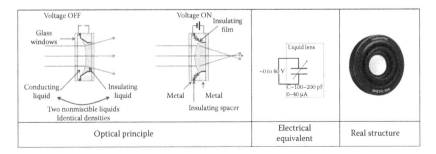

FIGURE 5.4 Optical schematics, electrical equivalent, and physical realization of the liquid lens.

(also called the wetting line or triple line), which is supported by a wall on the inner side of a conical hole. This conical wall is made of a metal piece covered with an insulating layer.

Following the application of a voltage, the wetting line moves reversibly along the cone. At zero voltage, the wetting line is at the cone's larger opening edge, whereas at maximum voltage the wetting line is at the cone's smaller opening edge. Due to the constant volume of the two liquids, the curvature of the spherical interface between the liquids changes with the voltage. This is the basis of the variable optical power of the liquid lens.

The basic principle relies on three key properties:

- Liquids: The two liquids are different in nature (one conducts electricity, the other is a nonpolar insulating medium) but they have the same density, in order to suppress gravitational effects and translational accelerations such as those present during a shock or a vibration.
- EW actuation: This enables the voltage to apply a force on the liquid's perimeter (the wetting line) in order to inflate the liquid's interface more or less. The accurate control of the wetting of fluids on the solid wall is needed for the lens to work properly.
- Drop (or meniscus) centering: The liquid lens is one optical component and it has to be combined with other optical components to form a larger optical system with several glass and plastic lenses. As usual in optical systems, the lens has to be kept strictly aligned with other optical components. Any variable lens has to ensure that it has control over its own optical axis.

5.2.2 Basic Performances of the Liquid Lens Based on EW: Large Range, Hysteresis Free

The large range of variation of optical power is one of the most beneficial characteristics of liquid lenses. This can be expressed in terms of either the power range or the phase shift ranges.

Figure 5.5 shows the variation of the optical power of a typical liquid lens as a function of voltage. Although most engineers are not familiar with diopters, this presentation is convenient, as diopters relate directly to inverse object distances (see Equation 5.1). Figure 5.5 shows that the liquid lens allows focusing down to a few centimeters object distance.

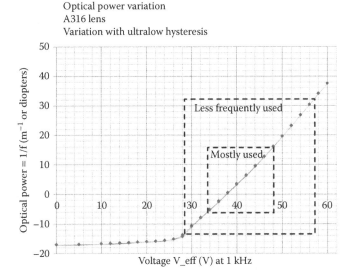

FIGURE 5.5 Focal response of the variable lens.

In terms of phase shifts, a liquid interface that is moving by large amounts, up to 0.5 mm for a 3 mm aperture liquid lens, realizes phase shifts up to several tens of waves. This is especially important for a transmissive adaptive optics element for focus adjustment: focus is one of the most demanding aberrations in terms of the range of phase shifts, especially at large diameters, and this is certainly one of the strong advantages of liquid lenses.

Another advantage of using liquid lenses is that the focus command is independent of the focal length of the fixed optics behind, unlike motors where the focus command (μm) converts to focus object distances through the focal length of the objective lens.

Also of note in Figure 5.5 is that there is virtually no hysteresis as it includes up and down variations over a complete cycle, and hysteresis is hardly measureable.

5.2.3 Wave-Front Error, Optical Quality

Since the introduction of digital cameras, tremendous progress has been made in sensor miniaturization and the quality of fixed optics, especially at large field angles. The specification of the wave-front error (WFE) of variable lenses will not be discussed in this chapter, as it depends on the complete camera architecture and on many other aspects, but typically the need is of the order of 50–100 nm for the maximum WFE over the useful pupil size (Simon et al. 2010). Figure 5.6 shows the WFE of a typical Arctic 316 lens over a measurement pupil of 2.5 mm diameter, on the same graph and in the same units as the focus variation.

The WFE is defined as the root mean square of the difference of the optical wave front from a perfect sphere. Figure 5.6 shows that the liquid lens has very large optical path variations in the focus mode, while keeping the residual errors very low in the other modes: the variation range is about 600 times the residual aberrations.

In absolute numbers, one sees that the particular lens shown here is not far from reaching the diffraction limit as defined by the Marechal criterion (WFE $\lesssim \lambda/15$). For some applications, such as laser beam manipulations, it is useful to know the decomposition of the residual aberrations of Zernike modes, as a way to better design an optical system. Figure 5.7 gives such a repartition,

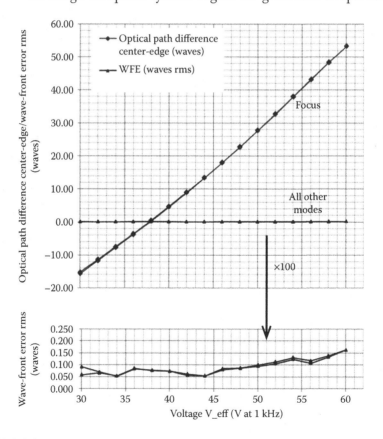

FIGURE 5.6 Focal response and residual aberrations, versus voltage.

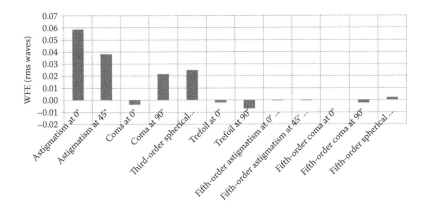

FIGURE 5.7 Typical pattern of the residual wave-front error decomposition with Zernike modes.

for the same typical lens representative of a standard production batch.

As expected, the weight of the aberrations concentrates on the first modes, rapidly disappearing to negligible values for higher Zernike modes.

5.2.4 Liquids Main Properties

The design of a working liquid lens relies on two nonmiscible liquids. It is of note that there is no membrane between the two liquids: they have to be sufficiently different in nature (polar vs. nonpolar) to ensure that their interface has enough interfacial tension at all operating temperatures.

An important characteristic of the liquids pair is their index difference, Δn. Typically, liquid formulations today are in the range of $\Delta n \sim 0.1$. This difference could appear small, as the usual indices in fixed lenses are in the range of about 0.5 (the difference between the index of glass and air). Nevertheless, the characteristic of the variable liquid lens is that the liquid interface experiences large displacements, thus producing large optical phase shifts.

Another important aspect concerns the density of the liquids: in order to function properly, the two liquids must have the same density while one is insulating and the other is electrically conducting, and while having different indices of refraction. Table 5.1 lists the main properties of a typical liquid used for liquid lenses.

TABLE 5.1 Typical Properties of Liquids Used in Liquid Lenses

	Conducting Liquid	Nonpolar Liquid
Index of refraction at 20°C: nD	1.384	1.498
Density (kg/m³) at 20°C	1.056	1.056
Conductivity (mS/cm)	2.5	<<< (not measurable)
Interfacial tension (mN/m) at 20°C		14.6
Viscosity (cS) at 20°C	5.3	2.7

5.2.5 Thermal Effects Linked with Liquids

The difference of coefficients of thermal expansion (CTE) of the two liquids plays no role in small lenses (below 2–3 mm diameter), but it has an important limitation on the temperature range of use of large lenses (>5 mm diameter) (Crassous et al. 2004). More precisely, the difference in the density of the two liquids increases linearly with the temperature, changing sign at the temperature at which density matching has been realized. Density matching can be easily achieved at a given temperature, but it is more difficult to attain over a wide temperature range.

The consequence of this temperature-induced density mismatch is that a coma aberration appears when the temperature departs from the temperature of density matching. Figure 5.8

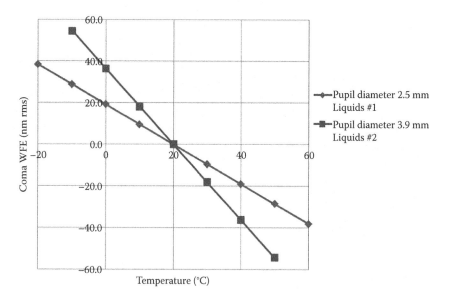

FIGURE 5.8 Coma aberration induced by thermal effects in liquid lenses.

shows the coma aberrations, expressed in WFE as a function of temperature, for two liquids that have been developed for two different lenses of different pupil diameters.

Ultimately, this is a limitation on lenses of large diameters: although practically feasible, a 20 mm lens based on EW could not tolerate the typical CTE mismatch shown in Figure 5.8, as it would require a precise temperature regulation. Nevertheless, a liquid's formulation is constantly progressing, and this limit is regularly surpassed by new liquids development.

5.2.6 EW: A Capacitive Actuation

For over 100 years, EW has been studied as a particular class of electrocapillarity. The pioneers of this field came from electrochemistry, such as Froumkine (1936) in the 1930s, who studied in detail the EW behavior of electrolytes at the surface of metals, against air or against an apolar liquid. Although large effects are induced by low voltages (typically a few hundred millivolts), the phenomenon is irreversible after a few cycles, due to surface staining and oxidation. Despite interesting trials to fix that problem by covering the conducting surface with some self-assembled monolayers (Sondag-Huethorst and Fokkink 1992; Lee et al. 1994), the system could not be made sufficiently stable to support a real application.

Things really changed when it became possible to block any current across the interface by using a thin (a micrometer or less) insulating layer. To our knowledge, this new layout was first released by N. Sheridon in the context of optical displays (Sheridon 1979) and other groups dealing with undermarine electrical cables (Minnema et al. 1980), before an understanding of the contact angle evolution with the different parameters involved had been reached. Over the last two decades, there has been considerable work on EW in different contexts, such as lab-on-chips and digital microfluidics (Mugele and Baret 2005; Quilliet and Berge 2001). The particular EW configuration used in liquid lenses is shown in Figures 5.9 and 5.10.

The key to using EW in an optical application is to suppress any source of solid friction that is often observed in wetting phenomena: any pinning of the triple line (oil/electrolyte/surface) produces defects that are visible as a degradation of the optical quality of the liquid lenses.

This has been the subject of extensive research on the supporting materials, as well as on the functional liquids. For instance, it

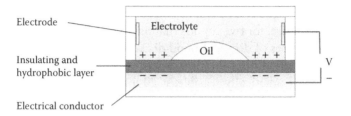

FIGURE 5.9 Schematic of the EW configuration used in the liquid lens.

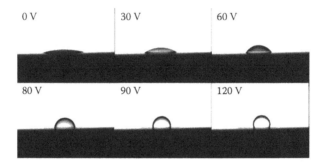

FIGURE 5.10 EW experiment according to sketch.

was shown that there is a direct relationship between the hysteresis of the contact angle (attributed to the pinning of the contact line) under EW, and the natural contact angle at 0 V (Maillard et al. 2009): the oil must cover the substrate, in order to form a lubricating layer underneath the electrolyte at any time during EW cycles, as shown in Figure 5.11.

5.2.7 Meniscus Centering

Another key technology that has to be included in the liquid lens answers the need for a stable optical axis for the variable lens. Of course, it is key to align the variable lens with other lenses in the system, so that the tolerancing of the optical overall system is kept inside the desired window.

In the liquid lens, this centering of the liquid's meniscus is achieved through the geometry of the supporting piece, which is conical (see Figure 5.4). This centering can be made by other means (for details, see Berge and Peseux 1997; Berge 1999), but for practical reasons the aforementioned is the easiest way.

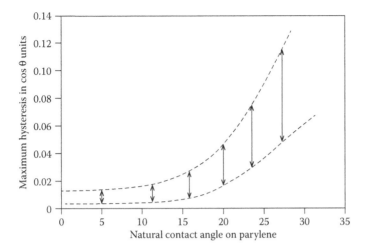

FIGURE 5.11 Maximum contact angle hysteresis measured using EW depending on the natural contact angle.

5.2.8 Variants of the Liquid Lens Architecture

Whereas the liquid lens architecture presented in this chapter is the only one to be industrially produced, several variants have been considered or prototyped in various locations. The Philips team has shown a cylinder type of lens, based on EW, using a piece of glass tube as external walls. Although not optimized for the optical interface curvature and angles, this structure has the merit of simplicity.

Another proposal has been made by a team in Singapore, where a couple of liquids were used at a fixed interface, the EW having been replaced by a pressure actuation. This architecture decouples the optical part from the pressure actuation. Nevertheless, the slow response time as well as the more complex architecture does not favor its industrial promotion.

More recently, Mugele and his team have advanced an interesting concept: the particular structure of this lens (see Figure 5.12) is that it enables collective motion through a pressure-induced EW actuation.

This list is not intended to be exhaustive. In particular, lenses made with a physical membrane and liquids are not mentioned, as they are more complex, their power consumption is higher due to the need to work against the membrane tension, and they are more fragile as they do not apply the principle of equal density of two liquids.

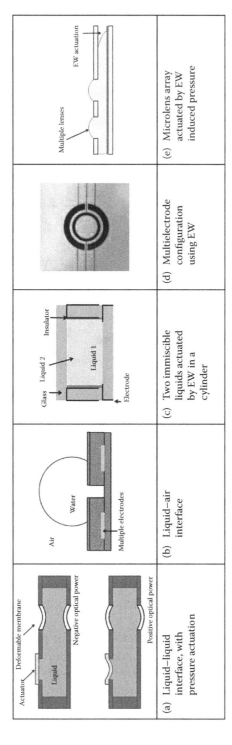

FIGURE 5.12 Various proposals of liquid lens architectures. ((a) Adapted from Moran, P., Dharmatilleke, S., Khaw, A., Tan, K., Chan, M., and Rodriguez, I., *Applied Physics Letters*, 88, 041120, 2006; (b) Adapted from Krupenkin, T, Yang, S., and Mach, P., *Applied Physics Letters*, 82, 316–318, 2003; (c) Adapted from Kuiper, S. and Hendriks, B. H. W., *Applied Physics Letters*, 85, 1128–1130, 2004; (d) Adapted from Park, J., A liquid lens based on electrowetting, PhD dissertation, Louisiana State University, 2007; (e) Adapted from Murade, C. U., van der Ende, D., and Mugele, F., *Optics Express*, 20, 18180–18187, 2012.)

5.3 Application of the Liquid Lens to AF Miniature Digital Cameras

5.3.1 Optical Architecture and Overall Dimensioning

The basic architecture of an AF digital camera using a variable lens is shown in Figure 5.13 (Saurei et al. 2004).

From top to bottom, or from the object side to the sensor, one finds:

- The liquid lens, which adjusts the focus of the camera. This lens is associated with its driver, usually a small chip to enable the voltage used by the liquid lens (low current).
- The "image-forming" lens, which is an association of three to five plastic or glass aspheric lenses that is designed to work at a fixed object distance (often infinity).
- An infrared (IR)-cut filter, which prevents the CMOS sensor from being blurred by near-IR rays that are invisible to the naked eye.
- The CMOS sensor itself.

Some manufacturers have placed the liquid lens at other positions in the optical system (inside the optical block, or at the image side of the main optics). The liquid lens is basically compatible with all these configurations, but this is more a matter of detailed optical design, which is not the purpose of this chapter.

In the standard configuration (a liquid lens is placed on the object side of the image-forming fixed lens), the relationship

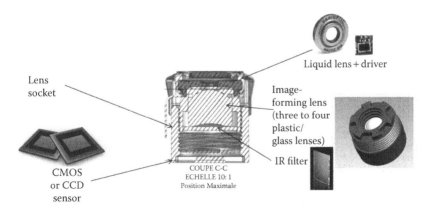

Liquid lens + driver

Lens socket

Image-forming lens (three to four plastic/ glass lenses)

CMOS or CCD sensor

COUPE C-C
ECHELLE 10: 1
Position Maximale

IR filter

FIGURE 5.13 Schematics of all elements composing a camera using a liquid lens.

between the object distance and the optical power variation of the liquid lens is given by

$$\frac{1}{d\_object} = \frac{1}{d\_object\_FixLensOnly} + \frac{1}{f\_LL} \quad (5.1)$$

where

d_object is the object distance from the camera

d_object_FixLensOnly is the object distance where the fixed image-forming lens has been adjusted to deliver a sharp image (when removing the liquid lens)

f_LL is the focal length of the liquid lens (inverse of the optical power)

Traditionally, it is easier to speak of inverse distance (m⁻¹) and optical power (m⁻¹ or diopters) as they are additives in some simple approximations. Figure 5.14 shows a typical situation with a depth of field of 0.3 m⁻¹ half width at half maximum (HWHM), typical of a 1/4-inch camera at f/2.8 aperture.

Considering the aperture of the liquid lens, the lens compatibility of a variable lens in cameras with current sensors is given in Table 5.2.

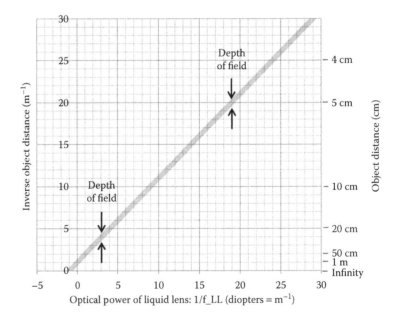

FIGURE 5.14 Relation between the optical power of the liquid lens and the object distance at which the camera delivers sharp images. The gray line gives the sharp zone, considering the depth of field of the system.

TABLE 5.2 Camera Compatibility Chart with Available Liquid Lenses

	Useful Aperture Diameter (mm)	Sensor Format (inch)	Focal Length (mm)	f24 × 36 Equivalent (Diagonal Field Angle) (mm)	f#	Range of Focus	Pixel Count
Baltic 617	1.6	1/4	3.7	35 (65°)	f/2.8	5 cm to infinity	All sensors (today's availability = 1 – >14 Mp)
		1/3.2	4.6	35 (65°)	f/2.8		
Arctic 316	2.5	1/3.2	4.6	37 (63°)	f/2.8		
		1/2.5	5.7	35 (65°)	f/2.8		
		1/1.8	9.6	44 (54°)	f/3.7		
A39N0	3.5	1/1.8	12	55 (45°)	f/3.1		
		2/3	16	60 (41°)	f/3.5		

Table 5.2 is only a base guide; of course, the detailed compatibility depends on f# of the image-forming fixed lens, the chief ray angles, and other parameters.

5.3.2 Performances of Camera Equipped with a Liquid Lens

Many aspects determine the overall quality of a camera equipped with a liquid lens. The two main parameters in the image quality are the performances of the CMOS sensor and the image-forming lens. Basically, the liquid lens does not drastically change the quality of the association between the fixed lens and the CMOS sensor. The liquid lens provides optical power to refocus the camera on the desired object, but the ray trajectories inside the fixed lens are not much affected by the vergence of the liquid lens.

This is also seen in the requirement for optomechanical accuracy when mounting the liquid lens on the fixed lens. The typical requirement of centering accuracy of an optical lens element depends mainly on its optical power: as the liquid lens is mainly used to correct focus, it will be a minor part of the main optical power of the objective, and thus it does not require the same accuracy of centering as the image-forming lens. The liquid lens will add about 10 diopters compared to the image-forming lens, which has an overall power of 200–300 diopters.

The same argument explains why the liquid lens, although having some chromatic dispersion, does not perturb the image formation: the chromatic dispersion of the lens cancels at 0 diopters, in the middle of the curve of variation of the liquid lens, thus the system is never far from the position of chromatic compensation.

Figures 5.15 and 5.16 demonstrate that the variable lens for AF miniature cameras can cover a broad range of object distances (typically 5 cm to infinity), without compromising the image quality.

5.3.3 Today's Main Market: Industrial Cameras

The variable lenses in cameras offer many different possible applications. Today, liquid lenses have started to be used in the industrial domain where highly stringent technical specifications are driving their use. VCMs present in consumer cameras cannot be present in industrial cameras due to their inherent fragility. On the contrary, liquid lenses, having equilibrated liquids (equal densities) and no moving parts, can serve in harsh environments. Several market segments have already started industrial products using liquid lenses:

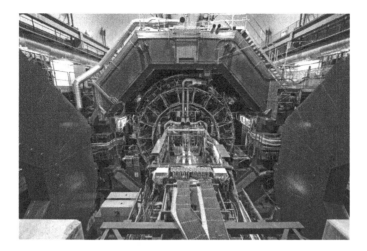

FIGURE 5.15 Images formed with the liquid lens for large object distances.

Bar-code reading: For a decade now, two-dimensional (2D) bar codes have benefitted from a strong tendency to replace one-dimensional (1D) bar codes: storing larger quantities of data (in particular, long URLs), with a large degree of redundancy for security, the 2D bar code is now well installed in many markets (medicine, industry, web-oriented apps, etc.).

FIGURE 5.16 Images formed with the liquid lens at short distances.

Contrary to the 1D bar codes, laser scanners cannot read 2D bar codes as easily as cameras do. Nevertheless, a sharp image is absolutely required in order to read newer bar codes. Here again the key is speed, but what counts is not just the overall speed of delivering a "good" reading of the bar code, but also the speed of capturing the image: one needs to take into account the accuracy of focusing because fast reading with lots of errors would be useless. The result is that image capturing with fast focus or fast AF has taken the lead and laser-reading systems have started to decline. The progress in image capturing and fast, reliable AF has been made by the variable lens architecture.

Biometry, security cameras: Here, the fast-focusing lens is key in order to benefit from the full resolution of the camera. A fixed focus could do the job but the user would need to adjust the distance of the camera to the target, which would result in slowing down the capture.

General purpose industrial cameras: The supply chain in modern industrial systems involves checking at many different stages the appearance of the product under construction, or the integrity of the parcel, or any other characteristics of the product that can be measured using a visual control. Figure 5.17 shows an example of such a camera.

Medical cameras: This broad range of cameras, ranging from invasive endoscopes to open-surgery cameras or dental cameras, all involve the use of a high-quality imaging system. Here, the challenge is to provide the doctor with an image

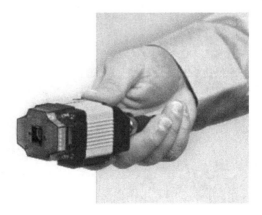

FIGURE 5.17 Industrial camera equipped with an EW-based liquid lens.

that is always sharp, which is necessary when performing complex surgical manipulations.

5.3.4 Potential for Future Applications

Table 5.3 lists the potential applications of liquid lenses based on EW. This table is not limited to focus-only liquid lenses, but it also includes multielectrode OIS liquid lenses (see Chapter 7) and also potential future products involving multiple electrodes. Indeed, a strong point of the liquid lens is that it scales easily from AF to AF + tilt and later to AF + tilt + astig, just by splitting one focusing electrode into four parts and later into eight parts. This is a strong advantage, especially compared to mechanical systems that require complex actuators and guiding systems to provide multiaxis actuation.

Table 5.3 shows that the technical improvements brought about by liquid lenses open many innovations for cameras, such as the ability to make long exposures in low light (10 lux or less), image stabilization during a flash-LED illumination, or in combination with super-resolution algorithms where the capability of inducing subpixel tilts could be a key advantage. Table 5.3 is clearly an indicative list of all applications of liquid lenses, which are unlimited.

5.4 Driving the Liquid Lens in a Camera

5.4.1 Electrical Properties and Power Consumption

The electric equivalent of a liquid lens is a variable capacitor: with increasing voltage, the capacitance increases as the cone area covered by oil decreases. Typically, the order of magnitude of the capacitance is about 200 pF, while the variation between no voltage and high voltage is less than 50 pF. The driving voltage for a given lens has to vary between 0 and V_max, and V_max ranges between 40 and 70 V, depending on the lens construction. Driving the currently available liquid lenses requires an ac driving signal, with the ac frequency varying from lens to lens, between one and several kilohertz (kHz).

The driving signals are usually provided by a small application-specific integrated circuit (ASIC) driver, including a step-up voltage converter and a full H-bridge to convert the dc output of the step-up in a fully symmetrical ac voltage source. The resulting

TABLE 5.3 Potential Applications of Liquid Lenses, versus the Optical Function Enabled by the Liquid Lens

Optical Features	Markets				
	Industrial Cameras	Consumer Cameras	Medical Devices	Ophthalmology	Other (Laser, Telecom, Pico Projectors, etc.)
Focus, autofocus Focus + tilt	Fast data capture Handshake correction	Fast AF video CAF Handshake correction	3D maps fast AF Handshake correction	Contact lenses IOLs	Projector focusing Fiber coupling multiple holography coding
Large range focus	Cheap or multiple microscopy	Magnifying glass microscopy	Confocal microscopy. Capsule endoscope	Super-vision. Low vision aid	Spherical aberration compensation (blue-ray)
Subpixels tilts	Super-resolution	Zoom (super-resolution)	Super-resolution		
Astigmatism fixed direction	Focus assistance	Focus assistance	Focus assistance		Focus assistance in reading heads. Laser diode beam shaping
8-electrodes astig with variable direction	Edge or stripes detection		Beam control	Eye testing instruments: focus and astigmatism correction	Fiber coupling multiple astigmatism compensation

power dissipation inside the lens is less than 0.5 mW, while the total power consumption including the driver reaches about 15 mW for AF lenses.

5.4.2 Driving Signals to Optimize Response Time

Many users have experienced that the best way to drive a liquid lens can involve reshaping the driving signal. For instance, when a step-by-step driving of the lens is required, it can be shown that one can accelerate the lens response at each step by replacing the direct driving signal with a signal that includes a transient overshoot for a few milliseconds. This is the case typically during the focus algorithm of an AF camera, which is sequenced by the frame periodicity.

Figure 5.18 shows an example of a ×1.6 acceleration of the step response of a liquid lens by adding a 10 ms-long transient overshoot of a given amplitude. The lower part of Figure 5.18 shows the driving signals as a continuous curve, on the left it is a simple step, on the right a transient overshoot has been added during 10 ms. The lower part of Figure 5.18 shows the evolution of the focal optical power of the lens as triangles, for both driving signals. This enables a first look at the response time of the liquid lens.

Nevertheless, looking only at the focal response is not accurate enough for some applications. Figure 5.18 (upper part) shows another monitoring of the liquid lens, through the WFE *including the defocus error*. This new measure is directly associated with the loss of the modulation transfer function (MTF) in a situation where the object is at a given distance (focus target), the camera evolving toward this given focus (shown in Figure 5.18 as squares). Figure 5.18 (upper part, left) demonstrates the response to the simple step, showing that the defocus error is slow and the residual WFE is fast: this leads to optimizing the response by increasing the voltage amplitude of the overshoot, in order to speed up the focus response while keeping the WFE lower than the specification.

To summarize Figure 5.18, applying an overshoot results in a decrease in the response time from about 40 to 25 ms, for this particular lens (older version of the liquid lens). Figure 5.18 shows the optimized amplitude of the overshoot as a function of the voltage step. As expected, the results show good agreement with a linear regime; indeed, the focal response of the liquid lens is known to be close to a second-order linear system.

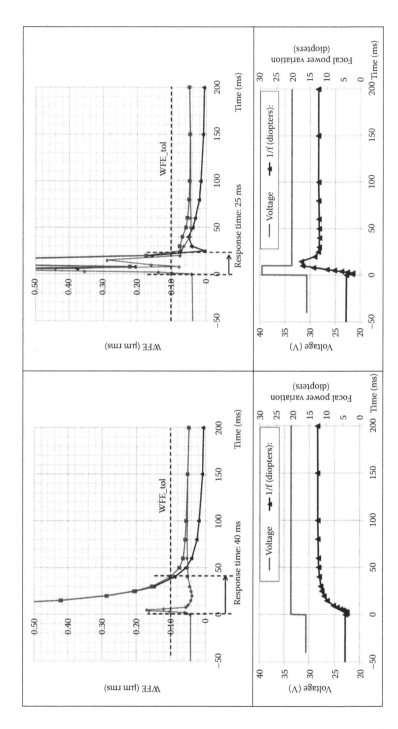

FIGURE 5.18 Upper part: Response curve for the wave-front error (diamonds), defocus (triangles), and WFE including the defocus (squares) in two different excitations. Upper left part is for a simple step +3 V. Upper right part is for an excitation with a short overshoot of +6 V amplitude during 10 ms. Lower part shows the voltage excitation (continuous) and the focus response (triangles).

The procedure described requires precise knowledge of the real operating aperture of the liquid lens. Once this is set, one can optimize the different parameters of the overshoot (time of the transient, amplitude, etc.), and benefit from a large acceleration factor, as shown in Figure 5.19, made with the same product reference as above.

One sees in Figure 5.20 that the acceleration factor is quite important, as the response time can move from above to below the frame period with the overshoot. In practical cases, there are several ways of installing this modification: either by software, adding a given calculated overshoot at each step, or by applying a linear filter. It has been shown that the amount of overshoot is roughly linear with the step size, enabling a high-order filter strategy. In some cases, the latter way is easier, especially when no clear periodic updates of the focusing mechanisms come from the system.

5.4.3 Ultrafast Algorithms for the Search of Focus

As was mentioned in Section 5.1, there are other ways to accelerate the AF mechanisms of cameras when using a liquid lens

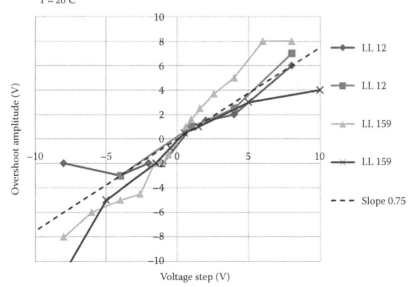

FIGURE 5.19 Amplitude of the overshoot (duration 10 ms) as a function of the step amplitude.

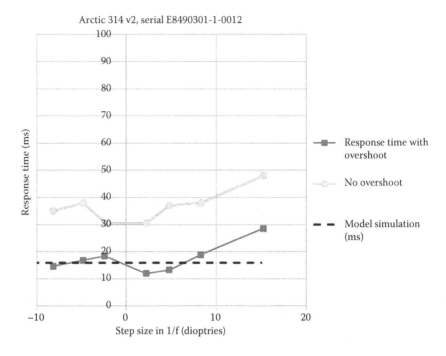

Arctic 314 v2, serial E8490301-1-0012

FIGURE 5.20 Result of the optimization of the response time by adding a transient overshoot to the driving signals: without overshoot (circles) the response time is of the order of 40 ms, whereas it is less than 20 ms with overshoot.

instead of a motorized system. The response of the liquid lens has been shown to be roughly independent of the step size, visible on the curve of Figure 5.17. Motorized systems run with a limitation on the maximum velocity, and thus have a settling time that is dependent upon the step size (see Chapter 2). With these systems, the natural tendency is then to use sequential algorithms, where the focus search is run step by step. In addition, motorized systems do not like reversing actuation, due to increased inertia and backlash.

A liquid lens, by offering random actuation to any desired focus position at the same speed, opens up the algorithm possibilities tremendously. Similar to RAMs, it is possible to considerably accelerate the search for the best focus by using nonsequential algorithms. Figure 5.3 shows several examples of such nonsequential algorithms, either by dichotomy (splitting the search range by ×2 at each step) or by splitting the search range by a factor of 0.62 at each step for the golden search. In this chapter, we do not examine these strategies in detail, but we mention these algorithms because they are particularly important for a broad range of imaging systems, from machine vision to consumer cameras. Faster algorithms to search for the best focus can bring an acceleration factor up to ×2 in search time, realizing a tremendous gain in ergonomy, or in production costs, or in vision interpretation reliability.

5.5 Conclusion

The liquid lens has caused an important revolution in miniature cameras, as slow and fragile motorized systems could be replaced by a rugged and no-moving-part variable lens system. Nowadays, this revolution has started in industrial cameras, whose technical specifications are highly demanding, such as professional barcode readers, machine vision systems, and medical cameras. As an AF actuator, liquid lenses have a very large range of power, with the capacity to rapidly move from macro to infinity. The technology is also ready to go broader, by adding some new functionalities, such as tilt and astigmatism.

References

Asatryan, K., Presnyakov, V., Tork, A., Zohrabyan, A., Bagramyan, A., and Galstian, T. (2010), Optical lens with electrically variable focus using an optically hidden dielectric structure, *Optics Express*, *18*(13), 13981–13992.

Berge, B. (1999), Drop centering device, U.S. Patent 7649692.

Berge, B. and Peseux, J. (1997), Lens with variable focus, U.S. Patent 6369954.

Berge, B. and Peseux, J. (2000), Variable focal lens controlled by an external voltage: An application of electrowetting, *European Physical Journal E*, *3*, 159–163.

Cheng, C. -C., Chang, C. A., and Yeh, J. A. (2006), Variable focus dielectric liquid droplet lens, *Optics Express*, *14*, 4101–4106.

Chou, Y. (2009), Camera module with piezoelectric actuator, U.S. Patent Application US2010/0271541 A1.

Commander, L. G., Day, S. E., and Selviah, D. R. (2000), Variable focal length microlenses, *Optical Communication*, *177*, 157–170.

Crassous, J., Gabay, C., Liogier, G., and Berge, B. (2004), Liquid lens based on electrowetting: A new adaptive component for imaging applications in consumer electronics, in W. Jiang and Y. Suzuki (eds), *Proceedings of SPIE 5639, Adaptive Optics and Applications III*, p. 143, SPIE Press, Bellingham, WA.

Froumkine, A. (1936), *Couche double, Electrocapillarité. Surtension, Actualités scientifiques et industrielles*, Hermann et Cie, Paris.

Graham, R. (1940), A variable focus lens and its uses, *Journal of the Society of America*, *30*, 560–563.

Hamada, Y., Fujii, T., Sakata, M., Nishio, Y., Tsujino, Y., Kuroki, K., and Kuwano, Y. (1988), U.S. Patent 4989958.

Hartmann, H. (1976), Method and apparatus for automatic focusing an optical system with a scanning grating, U.S. Patent 4048492.

Kawashima, Y. and Takahashi, T. (2007), Optical code scanner with automatic focusing, PCT patent application WO2009/061317 also U.S. Patent 2010/0294839.

Krupenkin, T., Yang, S., and Mach, P. (2003), Tunable liquid microlens, *Applied Physics Letters*, *82*, 316–318.

Kuiper, S. and Hendriks, B. H. W. (2004), Variable-focus liquid lens for miniature cameras, *Applied Physics Letters*, *85*, 1128–1130.

Lee, R., Carey, R., Biebuyck, H. A., and Whitesides, G. M. (1994), The wetting of monolayer films exposing ionizable acids and bases, *Langmuir*, *10*, 741–749.

Maillard, M., Legrand, J., and Berge, B. (2009), Two liquids wetting and low hysteresis electrowetting on dielectric applications, *Langmuir*, *25*, 6162–6167.

Minnema, L., Barneveld, H., and Rinkel, P. (1980), An investigation into the mechanism of water treeing in polyethylene high-voltage cables, *IEEE Transactions on Electrical Insulation*, *EI-15*, 461–467.

Moine, D., Gaton, H., and Berge, B. (2011), Ultrafast non sequential AF algorithms using liquid lens technology: An experimental study, Conference Paper, Imaging Systems and Applications, Toronto, Canada, July 10, 2011.

Moran, P., Dharmatilleke, S., Khaw, A., Tan, K., Chan, M., and Rodriguez, I. (2006), Fluidic lenses with variable focal length, *Applied Physics Letters*, *88*, 041120.

Mugele, F. and Baret, J. C. (2005), Electrowetting: From basics to applications, *Journal of Physics: Condensed Matter*, *17*, 705–774.

Murade, C. U., van der Ende, D., and Mugele, F. (2012), High speed adaptive liquid microlens array, *Optics Express*, *20*, 18180–18187.

Naumov, A. F., Loktev, M. Yu., Guralnik, I. R., and Vdovin, G. V. (1998), Liquid crystal adaptive lenses with modal control, *Optical Letters*, *23*, 992–994.

Park, J. (2007), A liquid lens based on electrowetting, PhD dissertation, Louisiana State University.

Press, W., Teukolsky, S. A., Vettering, W. T., and Flannery, B. P. (1992), *Numerical Recipes in C*, 2nd edn, Cambridge University Press, Cambridge.

Quilliet, C. and Berge, B. (2001), Electrowetting: A recent outbreak, *Current Opinion Colloid Interface Science*, *6*, 34–39.

Rudmann, H. and Rossi, M. (2004), Design and fabrication technologies for ultraviolet replicated micro-optics, *Optical Engineering, 43,* 2575–2582.

Saurei, L., Mathieu, G., and Berge, B. (2004), Design of an autofocus lens for VGA 1/4-in. CCD and CMOS sensors, *Proceedings of SPIE, 5249, Optical Design and Engineering,* pp. 288–296, SPIE Press, Bellingham, WA.

Sheridon, N. K. (1979), Electrocapillary imaging devices for display and data storage, *Xerox Disclosure Journal, 4,* 385–386.

Simon, E., Berge, B., Fillit, F., Gaton, H., Guillet, M., Jacques-Sermet, O., Laune, F., Legrand, J., Maillard, J., and Tallaran, N. (2010), Optical design rules of a camera module with a liquid lens and principle of command for AF and OIS functions, *Proceedings of SPIE 7849, Optical Design and Testing IV,* 784903, SPIE Press, Bellingham, WA.

Sondag-Huethorst, J. A. M. and Fokkink L. G. J. (1992), Potential-dependent wetting of octadecanethiol-modified polycrystalline gold electrodes, *Langmuir, 8,* 2560–2566.

Suzuki, T. (2010), Challenges of image-sensor development, *IEEE International Solid-State Circuits Conference Digest of Technical Papers (ISSCC),* pp. 27–30.

Treisman M. and Silver, J. (1988), Suspension system for a flexible optical membrane, U.S. Patent 4890903.

Wolterink, E. and Demeyer, K. (2010), WaferOptics mass volume production and reliability, *Proceedings of SPIE, 7716, Micro-Optics,* SPIE Press, Bellingham, WA, 771614.

Wu, H., Chou, C., and Shyiu, S. (2009), Voice coil driving auto-focus lens module, U.S. Patent 7612957.

CHAPTER **6**

Electrically Variable Liquid Crystal Lenses

Tigran V. Galstian

Contents

6.1 Introduction

The continued miniaturization of optical cameras creates a need for adaptive or variable optical elements that make up these cameras. Such elements may be used to relax the manufacturing tolerances of the camera, to increase its resistance versus the variations of environmental factors (e.g., temperature), or to improve its performance by introducing advanced functionalities, including

181

autofocus (AF), optical image stabilization (OIS), and optical zoom (OZ). The design specificities of optical cameras and the necessity of specific functions and various adaptive components are presented in other chapters of this book.

This chapter gives an overview of the use of liquid crystals (LCs) and the various geometries of electrically variable LC lenses (EVLCLs) for applications in optical cameras. However, it is important to justify our choice of EVLCL technology before discussing the design and performance details of optical cameras. Mobile cameras impose draconian requirements on the cost, size, performance, and reliability of their components. EVLCLs are based on very mature LC display (LCD) technology. This enables the simultaneous (wafer-level) fabrication of several thousand EVLCLs, which provides a unit cost that is significantly less than one dollar, simple scalability of volume production, and the promise of using a reflow process in the camera assembly. Indeed, their reliability is very similar to that of LCD products, which are proven to be very good (no moving parts). Their sizes may be very small; currently, the size of commercialized EVLCLs is $4.5 \times 4.5 \times 0.5$ mm. The next generation of EVLCLs will have their thicknesses reduced from 0.5 mm down to 0.36 mm and EVLCL prototypes have already been demonstrated with <0.28 mm thickness. This is quite comparable with the voice coil motor (VCM) travel distance, needed for focusing at 10 cm. EVLCLs may be driven by ac voltages (1 kHz) as low as 3 V root mean square (RMS). Their power consumption can be <10 µW and they are very light ($\approx$20 mg). Finally, they may combine AF and OIS functionalities in the same component. As will be seen in this chapter, their technical performances are quite comparable with other technologies; they have both advantages and drawbacks. However, I think that these factors are sufficient to warrant interest in EVLCL components.

6.1.1 Gradient-Index Lens

In this chapter, we will mainly focus on a family of electrically variable elements that is based on the dynamic generation of gradients of the optical refractive index $n(x,t)$; x is the radial coordinate for circularly symmetric lenses or the linear coordinate for cylindrical lenses. A well-known example of such an element is the gradient-index (GRIN) lens (see Figure 6.1; Moore 1977). Typically, these lenses are built with a constant thickness d and a spatially (mainly laterally) varying refractive index $n(x)$, in contrast with traditional

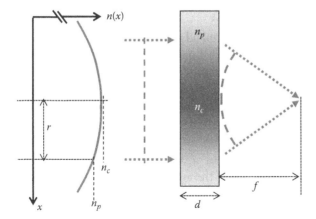

FIGURE 6.1 Schematic presentation of the refractive-index profile, $n(x)$, of a gradient-index lens; n_c and n_p are the central and peripheral refractive-index values; r, d, and f are the radius, thickness, and focal distance of the lens, respectively.

convex- or concave-shaped lenses. The careful control of the profile of $n(x)$ during the manufacturing process (e.g., by material diffusion; Moore 1980; Hensler 1975) enables obtaining the desired wave front for light traversing such components. Spherical or more complex (even W-shaped; Doyle and Galstian 2009) refractive-index profiles may be obtained and used to compensate for specific optical aberrations. The radial refractive-index profile of a standard GRIN lens may be described (Moore 1977) as

$$n(x) = n_c - a\,x^2 - b\,x^4 - c\,x^6 - \cdots \qquad (6.1)$$

where n_c is the value of the central refractive index and the coefficients a, b, and c define the profile of $n(x)$ in the transverse plane; $x = 0$ corresponds to the center of the lens. The optical power (OP), measured in diopters (D) or per meter^{-1} (m^{-1}), of such a component may be expressed (Goodman 1968) as

$$\mathrm{OP} = \left(\frac{1}{f}\right) \approx \frac{2d\left(n_c - n_p\right)}{r^2} \equiv \frac{2\Delta\varphi_0}{r^2} \qquad (6.2)$$

where
  n_p is the refractive-index value in the periphery of the lens
  r is the radius of the lens (half of the clear aperture CA)
  f is the lens focal distance (measured in meters) (Figure 6.1)

6.1.2 Electro-Optics of LCs

LC materials are well adapted to create electrically variable GRIN elements. These are liquids containing anisotropic molecules or molecular aggregates (Gennes and Prost 1995). The anisotropic polarizability of these molecules (typically having a rigid central part and flexible tails) and their close proximity (owing to the liquid state of matter) favor their natural parallel alignment (Gennes and Prost 1995). This, in turn, creates a local anisotropy axis that is often presented by a unit vector **n** (see Figure 6.2), called the "director," showing the average orientation of molecular axes. Note that, in the large majority of LCs, their physical properties are the same for the directions −**n** and +**n**. The anisotropy of LCs is at the origin of their confusing name (liquid *crystals*) since only crystals were known to possess anisotropic properties at the early stages of LC studies. Another key feature of LCs is the orientational flexibility of the director, which is due to the liquid state of the LC. Thus, one can easily reorient the director by applying different stimuli (see Equation 6.4). Interestingly, this angular deformation is "transferred" in space since consecutive molecular monolayers tend to maintain the orientation of their neighbors. Thus, if the orientation of the director is fixed in a given direction at a position z_0 (see Figure 6.3), then the tilt (reorientation) of the director in the vicinity of that point (at z_1) will generate some "elastic deformation"

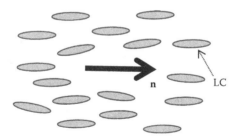

FIGURE 6.2 Schematic presentation of the natural alignment of LC molecules (filled ellipses) shown by the "director" **n**.

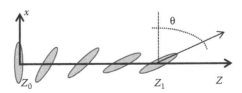

FIGURE 6.3 Schematic presentation of the "deformation" of LC's molecular orientation.

energy that would be proportional (via the so-called elastic constant K; typically at the order of 10^{-6} dyne) to the angle of the tilt θ and inversely proportional to the distance $d = z_1 - z_0$. Note also that the LC molecules will be oriented in a way to minimize the local gradient of the reorientation angle ($d\theta/dz$) between the points z_1 and z_0. In the simplest case of the in-plane "splay" deformation (the director remains in the x,z plane, see Figure 6.3), the additional elastic deformation energy may be expressed (Gennes and Prost 1995) as

$$F_d = 0.5K \left(\nabla \mathbf{n}\right)^2 \tag{6.3}$$

Thus, for a small angular deformation θ at the spatial scale d, the amount of elastic energy may be estimated as $F_d \sim 0.5\ K\theta^2/d^2$. To generate such a deformation, mechanical force and magnetic or electric fields may be used (Gennes and Prost 1995; Blinov and Chigrinov 1994).

The case of the electric-field-induced molecular reorientation is schematically demonstrated in Figure 6.4. The example used here is a molecule (filled ellipse) that has no spontaneous polarization (this model may be applied to LC materials composed of molecules without spontaneous dipoles or with a chaotic distribution of their dipoles). Thus, the dielectric torque here is obtained because of the *induced* dipole moment (dotted ellipse, Figure 6.4). Therefore, the value of the electric torque (applied to the molecule) may be expressed as

$$M_E \approx \Delta\varepsilon_m E^2 \sin a \, \cos a \tag{6.4}$$

where

E is the applied electric field

$\Delta\varepsilon_m$ is the anisotropy of the molecular polarizability

a is the angle between the electric field and the axis of the molecule (the director $\mathbf{n}$ in the case of the LC)

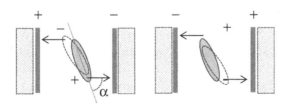

FIGURE 6.4 Schematic presentation of the dielectric torque applied by an ac field on a molecule (filled ellipse) via the induced dipole (dotted ellipse) interaction.

From this equation we can see that the torque, M_E, is zero for angles $a = 0°$ and $a = 90°$. However, for $\Delta\varepsilon_m > 0$, $a = 0°$ corresponds to a stable orientation state, while $a = 90°$ may be destabilized. Indeed, the thermal fluctuations of the director (small deviations from $a = 90°$) may increase the torque M_E. However, the corresponding (to the fluctuational deformation) elastic energy of the LC would try to reduce these fluctuations. The angular deviations may be increased if the electric field is strong enough (above a threshold, see Equation 6.6 [Gennes and Prost 1995; Blinov and Chigrinov 1994]).

This example also demonstrates the fact that the alternation of the electric field's polarity (shown in Figure 6.4 with the signs "+" and "−"), applied to avoid the macroscopic migration of ionic impurities, does not change the torque's direction, owing to the induced character of the dipole.

Given the fact that the orientations of the LC molecules are strongly correlated, the application of such a torque by using quite low electric fields (see Equation 6.6) appears to be enough to overcome the thermal fluctuation and elastic energy (Gennes and Prost 1995; Blinov and Chigrinov 1994). Thus, well-known consumer electronic applications, such as LCDs (Yeh and Gu 1999), and, more recently, lenses (Asatryan et al. 2010) have been developed using such electrical torque induced director reorientations. The operation of these devices may be explained by the structure and basic behavior of their LC cells. Thus, a typical cell of uniaxial LC (such as nematic LC [NLC]) is built by using two glass or polymer substrates (typically of 0.5 mm thickness). If needed, the external side of the substrate may be coated with broadband antireflection coatings to minimize light reflection (Fresnel) losses. The opposite (internal) side of the substrate is coated with a thin layer (≤20 nm) of a transparent electrode— indium tin oxide (ITO). Here also, multiple dielectric layers may be added to reduce the reflection losses. In this case, the layer is called "index-matched" ITO. The ITO layer may be patterned (e.g., nonuniformly etched) according to the use of the cell (see Section 6.2.1). A thin polymer layer (typically polyimide [PI], with ≤50 nm thickness) is then cast on the ITO layer by spin coating or printing. The mechanical rubbing of this last layer (after its thermal processing) provides a preferred orientation of the director when NLC molecules are put in contact with such a surface (Takatoh et al. 2005). Sandwich-like cells are then fabricated by using two such substrates (PI layers facing inside the cell)

(Figure 6.5). The injection of NLC materials into this sandwich (by capillarity after cell fabrication or by the "drop-fill" technique during cell fabrication) results in a uniform "monocrystal" (the director is the same over the entire volume of the cell) if the two rubbing directions (Figure 6.5, two vertical dashed arrows) are aligned in opposite directions (the so-called antiparallel alignment). In addition, PI layers usually provide (after rubbing) a small angle of the director's alignment θ_0 with respect to the surface of the cell (the *pretilt* angle; typically a few degrees), which is very important to obtain unidirectional reorientation (without spatially abrupt changes of orientation, called *disclination* defects) when an electric field E is applied to the NLC. Indeed, this field is usually applied with the help of two ITO electrodes of the sandwich, which generate an E field that is perpendicular to the substrate. Thus, because of the nonzero pretilt angle ($\theta_0 \neq 0$), all molecules of the cell are subjected to the same nonzero ($M_E \neq 0$) electrical torque (see Equation 6.4).

Light propagation in such an anisotropic medium allows the introduction of the differential phase delay $\Delta\varphi$ (Blinov and Chigrinov 1994) or of the optical path difference (OPD) between

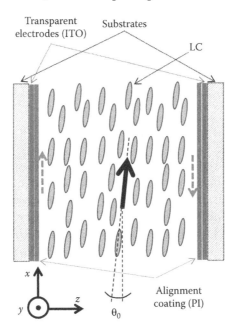

FIGURE 6.5 Schematic presentation of the LC cell in its ground state (with uniform director alignment) and its typical layers. The bold vertical arrow shows the director of the NLC. The two opposed dashed arrows show the PIs' rubbing directions.

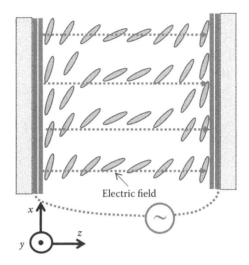

FIGURE 6.6 Schematic presentation of the molecular reorientation (owing to the dielectric torque applied by an ac field) that is nonuniform along the *z*-axis but is the same along the *x*- and *y*-axis.

the *ordinary* (the electric field of light is perpendicular to the drawing plane, which contains the NLC's director and the wave vector of light) and the *extraordinary* (the electric field is in the drawing plane) polarization modes of light, characterized by the ordinary n_o and the extraordinary n_e refractive indices, respectively. In the simplest case of normal light incidence on a cell without a pretilt angle ($\theta_0 \approx 0$), the value of $\Delta\varphi$ may be calculated as

$$\Delta\varphi(z) = \frac{2\pi z \Delta n}{\lambda_0} \qquad (6.5)$$

where λ_0 is the wavelength of light in vacuum and $\Delta n \equiv n_{\parallel} - n_{\perp}$ is the optical birefringence of the NLC ($n_{\parallel}$ and $n_{\perp}$ are the refractive-index values, respectively, for polarizations, parallel and perpendicular to the optical axis of the NLC). Thus, various polarization states may be obtained by the appropriate choice of the thickness of the NLC cell ($z = d$) and of its birefringence, Δn. In addition, because of the orientational flexibility of the director, the optical axes of the NLC may be reoriented by the application of an electric field (see Figure 6.6). This may thus change the local value of the effective birefringence $\Delta n_{\text{eff}}(z) \equiv n_e(z) - n_{\perp}$ and of $\Delta\varphi$ (see Equations 6.7 and 6.8). In reality, such modulation may be achieved if the applied electrical voltage is above the threshold value, defined (Blinov and Chigrinov 1994) as

$$V_{th}^2 = \frac{\pi^2 K}{(\varepsilon_0 \Delta \varepsilon)} \qquad (6.6)$$

where ε_0 is the dielectric constant of vacuum (8.85×10^{-12} C^2 m^{-2} N^{-1}) and $\Delta \varepsilon \equiv \varepsilon_{\parallel} - \varepsilon_{\perp}$ is the dielectric anisotropy of the NLC at the electrical driving frequency (typically an ac signal at 1 kHz). We can use typical values of the parameters of the NLC mixture TL 216 (from Merck ("Merk")) to estimate the corresponding threshold voltage value. Thus, we obtain $V_{th} \approx 1.76$ V by using $K = 15.3$ pN and $\Delta \varepsilon = 5.5$.

This torque forces the director of the NLC to be reoriented along the direction of the electric field for the case $\Delta \varepsilon > 0$. However, the alignment of molecules being fixed at internal surfaces of the cell substrates (by rubbed PI), the director's deformation is not uniform along the z-axis (Figure 6.6). Usually, for a relatively small reorientation and for infinitely strong anchoring at the cell surface (e.g., when the polar anchoring energy $W \gg 4Kd$), the tilt angle at the surface of the cell may be considered to be equal to the pretilt angle, θ_0, and the obtained spatial dependence of the tilt angle (between $z = 0$ and $z = d$) may be presented as $\theta(z) = \theta_0 + \theta_m \sin(\pi z/d)$, where θ_m is the maximum angle of the tilt, typically achieved in the middle of the cell. The value of the local birefringence, Δn_{eff}, will thus be defined by the local extraordinary refractive index, n_e, of the NLC, which depends upon the director's local tilt angle, $\vartheta(z)$, with respect to the wave vector of light:

$$n_e(\vartheta) = \frac{n_{\parallel} n_{\perp}}{\sqrt{n_{\perp}^2 \sin^2(\vartheta) + n_{\parallel}^2 \cos^2(\vartheta)}} \qquad (6.7)$$

Thus, the cumulative differential phase delay (between the ordinary and the extraordinary polarization modes of light) during the propagation along the z-axis will be defined as

$$\Delta \phi(z) = \left(\frac{2\pi}{\lambda_0}\right) \int_0^z \left(n_e(z) - n_{\parallel}\right) dz \qquad (6.8)$$

Among other important factors, we have to mention that the previous suggestion of a strong anchoring condition significantly influences the "off" time, τ_{off}, of the free relaxation of the director's orientation when the reorienting field is removed. This process is typically described as exponential decay, $\theta_m(t) \approx \theta_m(0)\exp(-t/\tau_{\text{off}})$. Thus, for the strong anchoring case, the time, τ_{off}, is defined by the rotational viscosity γ, the elastic constant K of the NLC, and its thickness d (Gennes and Prost 1995; Khoo and Wu 1993):

$$\tau_{\text{off}} \approx \frac{\gamma d^2}{\left(K\pi^2\right)} \tag{6.9}$$

We can use the typical values of the parameters of the same NLC mixture TL 216 to estimate the characteristic relaxation time of an NLC layer of thickness, $d = 50$ μm. Thus, by using the room temperature value of $\gamma \approx 297$ mPa s, we obtain $\tau_{\text{off}} \approx 4.9$ s. It is important to mention that the ratio γ/K depends upon the NLC's temperature; it increases at low temperatures (Wu et al. 1987). This is the reason the working temperature of the NLC devices (such as commercial LCDs, typically operating at $\approx 40°C$) must be increased either by warmer neighboring elements or by a microheater to reduce their response times. Second, as we shall see Equation 6.10, the EVLCL's performance may be significantly improved by the use of specific driving algorithms (a temporal sequence of electrical signals).

The value of τ_{off} is thus difficult to control once the geometrical and material parameters of the cell are fixed. In contrast, the excitation "on" time, τ_{on}, of the NLC's forced reorientation may be reduced by applying stronger excitation voltages (Blinov and Chigrinov 1994; Khoo and Wu 1993):

$$\tau_{\text{on}} = \frac{\tau_{\text{off}}}{\left(\left(V/V_{\text{th}}\right)^2 - 1\right)} \tag{6.10}$$

where V is the excitation voltage ($V > V_{\text{th}}$).

We can easily see that to obtain a high OPD (for a given NLC material), the NLC must possess high birefringence, Δn, and large cell thickness, d (Equation 6.5). However, we have to consider several tradeoffs since larger thicknesses will provide slower response times (larger τ_{off}, Equation 6.9). A higher elasticity constant K could help to reduce the τ_{off}, but it will also increase the threshold voltage, V_{th} (Equation 6.6). This is the reason much effort has been made to develop NLC mixtures with an optimized (mainly from the LCD's application point of view) figure of merit (FoM) (Wu et al. 1987):

$$FoM = \frac{(\Delta n)^2 K}{\gamma} \qquad (6.11)$$

In the case of the application of NLCs in imaging applications (e.g., in an EVLCL), the FoM must be reformulated since the high values of Δn are undesirable because of light scattering (see Section 6.2.3.3 [Gennes and Prost 1995]). Again, we face a trade-off since the reduction of Δn will reduce the maximum achievable OPD and the corresponding OP's variability range.

6.2 EVLC Lens

6.2.1 LC Lens Concepts

Research on EVLCLs started several decades ago (see, e.g., Bricot et al. 1977; Sato 1999). Many research groups were involved in the EVLCL development work. Among others, we should mention Akita University, Kent State University, University of Central Florida, University of Arizona, and Université Laval. The dominating approach was based on the GRIN lens concept, which requires a lenslike refractive-index modulation, described by Equation 6.1 (see also Figure 6.1). This type of profile may be achieved by gradually changing the NLC director's orientation when moving along the lateral x-direction of the lens (from its center, $x = 0$, to its periphery, $x = r$), as schematically shown in Figure 6.7. Many approaches have been developed to create appropriate gradients of electric fields, $E(x)$, that could create the required gradients of the director's orientation to obtain the needed $n_{\text{eff}}(x,z)$. Indeed, this task requires the careful choice of the profile of $E(x)$, taking into account the dependence of the refractive index, $n_e(x)$, versus the director's tilt angle, θ, and the balance of the dielectric torque, $M_E(x)$, with the deformation energy, $F_d(x)$.

The most straightforward approach (to generate the optimal profile of the reorienting electric field) would be to use multiple, individually addressed transparent electrode arrays (in linear or circularly symmetric forms, using a principle that is similar to that of the LCDs; see Figure 6.7 and Riza and Dejule 1994; Hashimoto 2009; Boss et al. 2011; Kato and Kawada 2007).

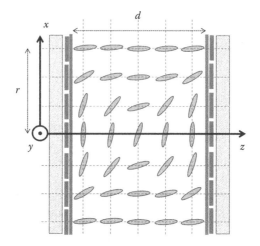

FIGURE 6.7 Schematic presentation of the lenslike molecular reorientation along the *x*-axis obtained because of the nonuniform dielectric torque applied by separate electrodes.

A cross section of these electrodes is shown by two vertical, bold, dashed lines between the PI layers and the cell substrates.

We shall now consider the case of linear electrode arrays (e.g., for a cylindrical lens) in more detail since this example could help us to understand other EVLCL solutions too. Let us suppose that we have a planar-oriented (in ground state) NLC with positive optical and low-frequency dielectric anisotropies ($\Delta n > 0$, and $\Delta \varepsilon > 0$ measured at 1 kHz). In this case, to generate a positive EVLCL, we must apply a high voltage value, V_0, at the peripheral pixels (at $x \approx \pm r$, see Figure 6.7) and continuously decreasing voltage values ($V_{m+1} < V_m$, where m is the distance of the pixel from the periphery of the lens) for pixels closer to the center of the lens (at $x \approx 0$). In this case, the extraordinary polarized light ray, which is propagating along the *z*-direction in the *central* area of the lens ($x = 0$, where there is almost no director perturbation), will undergo a phase delay:

$$\Delta \varphi_c \left(x = 0, z = d \right) \approx \frac{2\pi d n_{\parallel}}{\lambda_0} \tag{6.12}$$

In contrast, the same extraordinary ray, which is propagating at the *periphery* of the lens ($x = \pm r$, where the director is almost perpendicular to the substrates due to the strong, local voltage value V_0), will undergo a different phase delay:

$$\Delta\varphi_p\left(x=r, z=d\right) \approx \frac{2\pi d n_\perp}{\lambda_0} \tag{6.13}$$

thereby generating an OPD:

$$\Delta\varphi = \Delta\varphi_c - \Delta\varphi_p \approx \frac{2\pi d \Delta n}{\lambda_0} = \frac{\Delta\varphi_0 2\pi}{\lambda_0} \tag{6.14}$$

which is necessary for obtaining nonzero OP of the GRIN lens (see Equation 6.2).

Here also, we can use the typical values of the parameters of the NLC mixture TL 216 to estimate the theoretically achievable maximum OP level by using Equation 6.2. Given that, in the visible range, we have $n_\perp = 1.5237$ and $n_\parallel = 1.7349$, then, to achieve 10 D of OP for an EVLCL with CA = 1.5 mm, we would need (see Equation 6.2) an NLC layer with thickness $d \approx 10 \text{ m}^{-1} \times (1.5 \times 10^{-3} \text{ m})^2/(2 \times 0.21) \approx 53.6 \text{ μm}$.

It is important to mention here that the described molecular reorientation mainly happens in the plane of drawing. Thus, such an NLC layer would allow only the focusing of extraordinary polarized light since ordinary polarized light will "see" the same refractive index, $n_\perp$, everywhere. However, natural light (obtained from the sun, lamps, etc.) is not polarized and may be represented as a combination of two noncoherent, cross-oriented, linear polarized light components. Thus, only half of that light will be focused. This is the reason we can call the aforementioned lens a "half" EVLCL. To image objects in natural light, we would need two NLC layers to build a "full" EVLCL that would be capable of focusing both polarization components of light. This kind of "full-lens" may be obtained, for example, by building two identical "half-lenses," then rotating one of them by 90° and joining them (see Figure 6.8). Thus, during the lens operation, the NLC molecules of the first layer (left cell with $LC_\parallel$, Figure 6.8) are reoriented in the lenslike form with their axes remaining in the x,z plane (focusing light propagating along the z-axis with its linear polarization in the x,z plane), while the molecules of the second NLC layer (right cell with $LC_\perp$, Figure 6.8) are reoriented in the lenslike form but with their axes remaining in the y,z plane (focusing light with polarization that is in the y,z plane). The distance, Δz, between these two NLC layers must be as small as possible to minimize the difference in

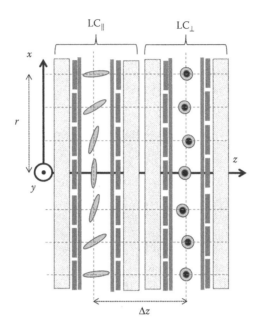

FIGURE 6.8 Schematic presentation of the combination of two "half" lenses into a polarization independent "full" lens.

the focusing positions of two cross-polarized light components. Also, for the same reasons, the OPs of two half lenses must be identical. It is interesting to note that this type of "polarization aberration" is probably negligible in classical lenses, which are built using isotropic materials. In the future, we shall keep this necessity (of using full lenses) in mind when considering other types of EVLCLs. However, we shall mainly describe the half-lens geometries for the simplicity of introducing their operation principles.

As we have already mentioned, the multielectrode approach introduced earlier is very straightforward. However, several practical considerations limit the application of arrayed and individually addressable multiple electrode EVLCLs. Among these, we can mention the cost of manufacturing, the complexity of the dynamic control, and the cost of the application-specific integrated circuit (ASIC) that is used for controlling the EVLCL.

The spatially nonuniform (lenslike) refractive-index profile, $n(x)$, may also be generated by using only two uniform ITO electrodes (1 and 2, Figure 6.9) and a nonuniform, polymer-stabilized LC (PSLC) (Nose et al. 1997; Presnyakov et al. 2002; Ren and

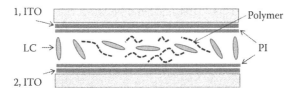

FIGURE 6.9 Schematic presentation of a polymer-stabilized LC lens.

Wu 2003). In this case, the electric field is relatively uniform across the diameter of the lens, and the nonuniform (lenslike) molecular orientation is obtained owing to the nonuniform morphology (density, etc.) of the polymer network that is dispersed in the LC matrix (Figure 6.9). The close positioning of two electrodes here enables a low-voltage operation. The use of flat uniform electrodes is another advantage. However, the instability of the material morphology and the relatively high light scattering are among the key drawbacks of this approach.

The advantage of two flat uniform electrodes is also used in other EVLCL designs, which do not use PSLCs. Thus, an *external* fixed lens is used as a tool to obtain a spatially nonuniform dielectric "voltage divider," similar to the well-known electronic circuits with serial resistances (Wang et al. 2004b). Indeed, the required lens-shaped voltage drop (and the corresponding electric field that is applied to the NLC layer) is achieved by inserting a concave- or convex-shaped fixed lens (made of glass) between the two uniform ITO electrodes (see Figure 6.10). Because, at driving frequencies (at the order of 1 kHz), the dielectric constant ε_L (typically < 7) of the fixed lens structure is higher than the dielectric constant of air ε_A (equal to 1), this will generate a lens-like voltage drop for different points across the diameter of the EVLCL. Thus, the NLC layer will be subjected to a lens-shaped electric field, which is stronger at the periphery compared to the

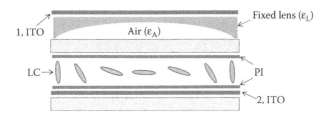

FIGURE 6.10 Schematic presentation of an LC lens using an external air gap and a fixed lenslike structure, with different dielectric constants $\varepsilon_A \neq \varepsilon_L$.

center of the lens. This happens because the drop of the electrical potential on the thick air gap is stronger and thus a weaker electric field is applied on the NLC layer in the center of the lens. The director's reorientation is correspondingly nonuniform. This approach, however, requires an increased distance between ITO electrodes 1 and 2 and provides an electrically uncontrollable or, a so-called, residual OP (ROP), both being caused by the fixed lens structure. The ROP here may be eliminated by using an optical material instead of the air gap (Asatryan et al. 2010). The refractive index of this material, n_{IM}, may be chosen to be equal to the refractive index of the fixed lens structure, n_L (typically ≈ 1.5). Then, the fixed lens structure is optically hidden and the ROP disappears (ROP = 0 for $n_{IM} = n_L$). This also eliminates Fresnel losses from the internal surfaces of the lens structure (Figure 6.10). In addition, several optical materials have relatively high dielectric constants ($\varepsilon_{IM} \approx 50$) at driving frequencies (at 1 kHz), which may noticeably increase the contrast of dielectric constants (ε_L vs. ε_{IM}), further improving the modulation capacity of the electric-field profile. However, as we have already mentioned, the insertion of the fixed lens structure imposes larger distances between two ITO electrodes, which, in turn, increases the driving voltages (several tens of volts). Thus, the high voltages and the required manufacturing precisions remain important drawbacks of this approach.

Curved structures have been used in several other approaches. One such approach still uses flat uniform electrodes, which are combined with a fixed lenslike structure that is, however, placed *inside* the NLC layer (Figure 6.11) (Sato 1979; Wang et al. 2004a; Knittel et al. 2005). The operation of this lens does not use a lenslike molecular orientation profile (so this is not a GRIN design). Indeed, the application of a relatively uniform electric field generates a lenslike OPD owing to the nonuniform

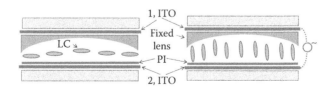

FIGURE 6.11 Schematic presentation of a lens using LC with nonuniform thickness owing to the fixed lenslike internal structure. Ground (left) and excited (right) states are shown.

thickness, $d(x)$, of the NLC (which is larger in the center than at the periphery of the EVLCL, $d_c > d_p$) and to the difference between the refractive-index values of the lenslike structure, n_L, and of the NLC (between $n_{||}$ and $n_\perp$, see below). Thus, let us consider an NLC cell where the ground state (no voltage) alignment of the director is planar (the director is parallel to the cell substrates, left side of Figure 6.11). The application of a sufficiently strong field may generate a perpendicular (to cell substrates) orientation of the NLC molecules (also called "homeotropic" alignment, right side of Figure 6.11). Thus, if the choice of NLC and fixed-lens materials is made in such a way that the refractive index of the lenslike material, n_L, and the ordinary refractive index of the NLC $n_\perp$ is the same (that is $n_L = n_\perp$), then, in the excited state (right side of Figure 6.11), and for the normal light incidence, the entire structure will appear as a uniform optical window without focusing properties. In contrast, when the E field is removed (and thus the NLC is relaxed back to the ground state), the normally incident extraordinary polarized light will "see" the refractive index $n_{||}$ of the NLC, which is different from n_L. Then, the lens structure will be revealed. The corresponding OP's value (for a given radius R of the fixed lens structure) may then be estimated as $OP \approx \delta n / R$, where $\delta n \equiv n_{||} - n_L \neq 0$, (Hecht 2001).

Another example of the use of curved structures is the spherically shaped external electrode (1) in combination with a uniform electrode (2) (Figure 6.12). In contrast with the previous case, here a uniform NLC layer is used, which is easier to manufacture. A fixed lenslike external structure (made of glass or polymer) supports a correspondingly spherically shaped external curved electrode (Wang et al. 2002). Thus, the lenslike electric-field profile is obtained by the control of the distance of two electrodes 1 and 2 (closer in the periphery). However, in this case also the EVLCL has an ROP. This ROP may be eliminated, e.g., if the

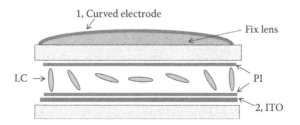

FIGURE 6.12 Schematic presentation of an LC lens using an external curved electrode.

fixed lenslike structure is covered with a similar material (with the same refractive index) having a flat top surface (Ren et al. 2004). However, the manufacturing complexities and the high voltages required for its operation reduce the area of practical use of such solutions.

Another interesting approach to obtain the lenslike electric-field profile was developed by using a pair of electrodes, one (1) had a hole in the center (Figure 6.13), while the other (2) was uniform (e.g., ITO) (Ye et al. 2004; Ye and Sato 2002). In this case, the electric field's gradient may be shaped to obtain a soft, lenslike refractive-index profile if the hole-patterned electrode (HPE) is positioned relatively far from the uniform electrode. In fact, the uniform electrode is usually positioned close to the NLC layer, while the HPE may be positioned on the internal (see Figure 6.14) or external surface of the top substrate (Figure 6.13). The position of the HPE is important since there is an optimal ratio between the diameter D of the lens and the cell thickness d: $D/d \approx 2$–3 (Sato 1999). Thus, this approach would only allow the fabrication of lenses of micrometric diameters (CA < 150 µm) if the HPE is positioned inside the NLC cells with typical thicknesses of $d \approx 50$ µm. Consequently, the fabrication of EVLCLs with millimetric CAs requires larger electrode distances (the HPE being placed on the external side of the NLC cell, Figure 6.13) and thus relatively high driving voltages (several tens of volts).

Another drawback worth mentioning, inherent in almost all EVLCL solutions (unless specific measures are applied to avoid it), is the formation of dynamic disclinations when an electric potential difference is suddenly (abruptly) applied to two control electrodes of the lens. In this particular case, the disclination problem is resolved in a configuration where a third uniform (independently controllable) electrode is added on the top of the HPE (Ye et al. 2004). However, the high voltages are still a

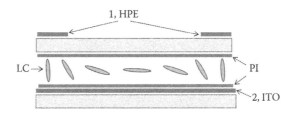

FIGURE 6.13 Schematic presentation of a lens using an "external" hole-patterned electrode (HPE).

significant problem of this geometry. In addition, the complexity of manipulating two independent voltage values increases the cost of the ASIC driver.

The problem of high voltages is resolved in the design where the HPE is positioned inside the NLC layer (see Figure 6.14; Naumov et al. 1998). In this case, for typical lens parameters, we have $D/d \approx 57$. Therefore, an additional high resistivity or a weakly conductive layer (WCL) must be cast over the HPE to soften the electric field's profile inside the NLC layer. Here, the electric field's spatial form is governed by the spatial attenuation of the electrical potential $U(x)$ (along the substrate plane, when moving from the border of the HPE to its center; Figure 6.14). This process is similar to the attenuation of electrical signals taking place in classical chains of electronic resistor–capacitor (RC) circuits. Indeed, each Δx_i slice (Figure 6.14) of the EVLCL may be characterized by a unit area capacitance, C_i (the NLC material being placed between two electrodes as a dielectric), and by an electrical resistivity, R_i, of the WCL. Thus, if the material and geometrical parameters of the EVLCL are appropriately chosen, then the attenuation of $U(x)$ will be optimal; strong enough to have the required electric field's lenslike profile, but not too strong to avoid its abrupt changes and associated with it optical aberrations (see Section 6.2.2.2 and Figure 6.21).

There are, however, important drawbacks with this approach too. Namely, most miniature cameras typically require EVLCLs with CA values ranging from 1.5 to 3 mm. The values of the dielectric constants of the NLC (ε_{LC}) typically range between $\varepsilon_\perp \approx 4$ and $\varepsilon_\parallel \approx 15$ and the thicknesses d of the NLC layers are between 20 and 50 µm. In this case, the "RC factor" of such lenses requires sheet resistance Rs values of the WCL that are in the range of several tens of MΩ/□. The fabrication of such films is extremely difficult since those sheet resistances correspond to the

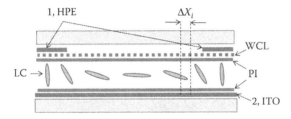

FIGURE 6.14 Schematic presentation of an LC lens using an "internal" hole-patterned electrode (HPE) and a weakly conductive layer (WCL).

transition (often called "percolation") zone between the dielectrics and the conductors. This significantly reduces the part-to-part reproducibility of lenses, which is particularly severe in the case of full lenses since two *identical* WCLs must be used. In addition, many types of thin films, with appropriate values of Rs, appear to be unstable (not reliable). Furthermore, even if two WCLs are identical, the refractive-index profile noticeably changes with the change in the OP level (since the unit area capacitance, C_i, may change by a factor of ≈ 3 by the rotation of molecules). This may bring a noticeable reduction of the modulation transfer function (MTF) of the camera using such tunable lenses (see Section 6.2.3.1 and Figure 6.25). This problem may be partially resolved (Naumov et al. 1999) by simultaneously changing the voltage and the frequency of the driving electrical signal, but this greatly complicates (and makes more costly) the corresponding ASIC driver.

Ye et al. (2010) proposed a similar (to the previous geometry) approach. In this case, an additional disc-shaped electrode (DSE) is positioned in the middle of the HPE (Figure 6.15). The independent control of its voltage allows partial control of the refractive-index profile and enables obtaining both positive and negative OPs (and also reduces the dynamic appearance of disclinations). However, the same problems of reproducibility, driver cost, and MTF degradation also make this approach unattractive.

The problem of WCLs reproducibility is resolved in the EVLCL geometry, shown in Figure 6.16 (Galstian et al. 2011). Indeed, the specific (symmetric) geometrical position of the HPE and the WCL allows its use for the simultaneous control of two NLC layers, one with a director that is parallel to the drawing plane ($LC_{\parallel}$, Figure 6.16) and the other with its director in the perpendicular plane ($LC_{\perp}$, Figure 6.16). Thus, both polarizations of incident light may be synchronously focused here.

In addition, the use of the modern techniques of thin glass manipulation allows a significant reduction of the driving voltages.

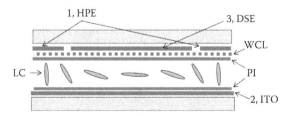

FIGURE 6.15 Schematics of a lens with two independently variable voltages.

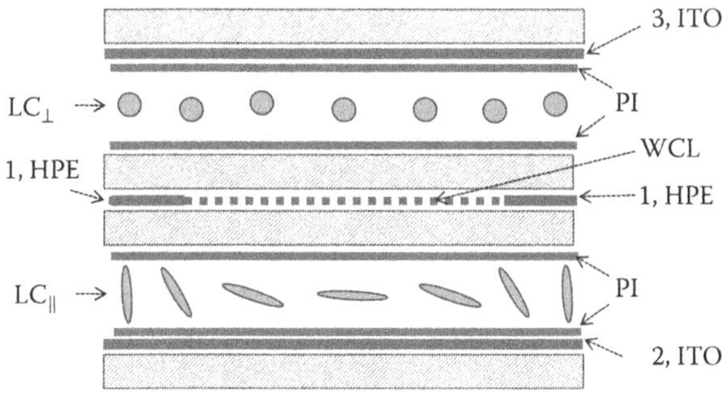

FIGURE 6.16 Schematics of a full (polarization independent) lens using the same HPE and WCL for the simultaneous control of two cross-oriented NLC layers ($LC_{\parallel}$ and $LC_{\perp}$).

This EVLCL geometry is also less sensitive to NLC reorientation owing to the presence of intermediate glass substrates. In addition, different electric-field profiles (and thus OP levels) are obtained for different driving frequencies at fixed voltages because of the resonant character of the effective chain of the $R_i C_i$ circuit of the EVLCL. Thus, a very cost-effective ASIC driver is used here with a simple frequency control of the OP level (at fixed voltage).

There are many other (more or less "exotic") approaches where LCs may be used to build electrically variable lenses, including the use of magnetic or thermal stimuli (Galstian et al. 2012a,b) and the lenslike gradients of the director's pretilt angle (Galstian 2010; Tseng et al. 2011).

A distinctive group represents the Fresnel lenses (Sato et al. 1985; Valley et al. 2010; Fan et al. 2003), which operate on the diffraction principle. While such lenses could be used for relatively large CAs (up to square centimeters), they remain costly and without the capacity to gradually change their OP level.

Another example is the use of LCs for the fabrication of lenses with micrometric CAs (Reznikov et al. 2012; Commander et al. 2000; Lee et al. 2011), which is not within the scope of this book. New EVLCL designs and applications are continuously being developed.

6.2.2 Technical Performance of LC Lenses

Hereafter, we shall limit our consideration to a specific family of EVLCLs, which was recently developed and commercialized by LensVector. These lenses are fabricated in wafer form; an example of such a wafer is shown in Figure 6.17.

The operation of this EVLCL is greatly simplified by the introduction of a purely frequency controlled technique (to change the

FIGURE 6.17 Example of the wafer manufacturing of EVLCLs. (From LensVector. With permission.)

OP at a fixed voltage), which is preferred to the traditional voltage control. An example of a transfer function (OP vs. driving frequency) is presented in Figure 6.18 for a lens with CA = 1.8 mm. As we can see, the OP increases very smoothly with the increase in the driving frequency. In fact, this is an analog device and its control granularity is defined only by the steps of the control frequency. Note that the 10 D OP level (e.g., needed for bar-code reading in mobile-phone or tablet cameras) is achieved at approximately 37 kHz. It is important to mention that the OP of the lens has a nonmonotonic behavior. Indeed, after the above-mentioned growth period, it

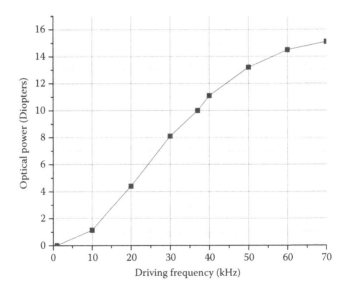

FIGURE 6.18 Dependence of the LC lens optical power upon the driving frequency (voltage is fixed).

decreases with the further increase of the driving frequency (this decreasing part is not shown here since the recommended working zone is on the growing slope, before achieving the maximum OP). The corresponding wave-front RMS aberrations (in micrometers) of the lens are continuously growing along with the increase in OP (Figure 6.19). As one can see, the RMS aberrations are at the order of 0.1 μm when the lens reaches 10 D OP at 37 kHz of driving frequency (shown by a vertical dashed line).

6.2.2.1 Speed of Transitions between OP Levels

NLCs are commonly considered as relatively slow media. However, as we can see in Section 6.2.3.2 and Figure 6.26, the AF convergence time (AFCT) of cameras, using EVLCLs, may be very short. It is worth mentioning that the speed of the change of OP is different depending upon the "direction" of the change. For example, let us consider the case when the NLC layer is planar oriented (in the ground state) and the low OP level (e.g., zero diopters) corresponds to the situation when almost all molecules are pointing perpendicular to the substrates (homeotropic or excited state). Then, changing the OP from 0 D state toward the high OP level would require the partial orientational relaxation of molecules in the center of the EVLCL. This will be the "slow" transition direction (typically several hundreds of milliseconds).

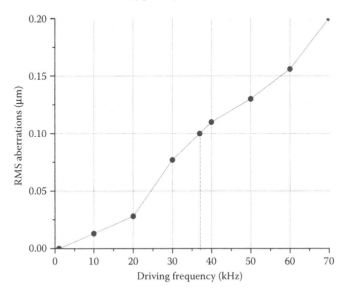

FIGURE 6.19 Dependence of the RMS aberrations of the LC lens upon the driving frequency (voltage is fixed). The dashed vertical line (at 37 kHz) corresponds to 10 D of OP.

In contrast, the change of OP from high to low levels suggests the application of an electric field that reorients (excites) all molecules toward the perpendicular state. This is the "fast" transition direction, which is typically at the order of 100 ms and can be further accelerated if the applied electric field is stronger. Figure 6.20 schematically shows that the fast and slow transition times may differ by a factor of 5, depending upon the excitation and relaxation conditions. As one can also see, the OP transitions are monotonic, without oscillations, jitter, or ringing effects. It is also important to emphasize that they have neither gravity dependence nor OP hysteresis versus the driving frequency.

As we have already mentioned, the speed of molecular rotation is defined by the ratio γ/K, which may be significantly improved by the appropriate choice of the "working temperature" of the lens. In addition, the use of specific driving techniques such as "overdrives" and "undershoots" significantly reduces the transition times (also, see the results of using specific AF algorithms in the following sections).

6.2.2.2 Refractive-Index Profile

For almost all imaging applications, the wave-front (or phase) shape of light, crossing the EVLCL, is a key parameter. Very

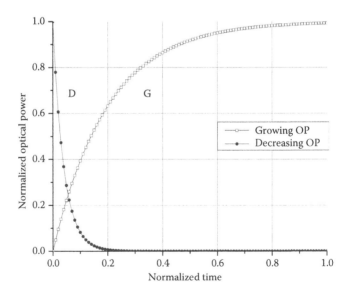

FIGURE 6.20 Schematic demonstration of the transitions speed difference when the optical power of the lens decreases (D) and grows (G).

often (not always!), this wave front should be spherical for all OP values. In reality, however, almost all EVLCL technologies are altering this form when the OP level is changed. Several physical parameters of the EVLCL can affect the gradient of the electric field at different positions across the CA of the lens, from the periphery to its center. One typical decreasing function is the exponential function. Another is an attenuation law that may be expressed by Bessel functions (Vdovin et al. 1999). Figure 6.21 schematically shows three different cases with low (L), medium (M), and high (H) attenuation coefficients for the exponential attenuation case. For comparison, we also present the typical case of Bessel-type attenuation (B) and the closest quadratic (Q) form. As one can see, the Bessel-type attenuation provides a flat periphery and a flat center of the wave-front modulation. The "simple" exponential decay law will often provide a wave front with a flat periphery (curve L) or a flat center (curve H) and will allow only a rough approximation (curve M) of the desired quadratic wave front (curve Q). And often, this approximation level changes with the change in the OP of the lens. This, in turn, degrades the MTF performance of the lens (see Section 6.2.3.1 and Figure 6.25).

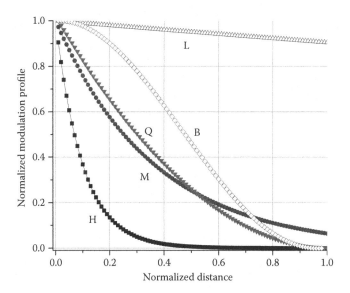

FIGURE 6.21 Schematic demonstration of the different modulation profiles of the phase front, from the periphery of the lens, at 0, to the center of the lens, at 1 (see text for details).

FIGURE 6.22 Wafer-scale autofocus camera using EVLCL. (From LensVector. With permission.)

6.2.3 Camera Performance Using LC Lenses

The wafer-scale manufacturing of EVLCLs enabled the manufacturing of the first wafer-scale AF camera (see Figure 6.22).

The quality of the images, recorded by cameras using EVLCLs, is quite good. Figure 6.23 shows an example of a picture recorded with low OP of the EVLCL (OP ≈ 0 D). The ISO chart and the flowers are positioned at approximately 1.5 m distance from the camera. Therefore, they are in-focus while the bill on the right side of the scene is out of focus since it is positioned much closer to the camera, at approximately 0.1 m. Figure 6.24 shows the same scene, but the OP of the EVLCL is now increased up to 10 D. As we can see, the bill now appears much clearer while the ISO target and the flowers look blurry.

FIGURE 6.23 Focusing on far objects. (From LensVector. With permission.)

FIGURE 6.24 Focusing on near objects. (From LensVector. With permission.)

6.2.3.1 MTF

Comparative studies of the MTF have been performed (at Nyquist/4) using an EVLCL that was integrated to a 1/4-inch camera with an F/2.8 and 5 megapixel complementary metal-oxide-semiconductor (CMOS) sensor. MTF data were gathered across the full field of view (approximately at a 60° diagonal), averaged over many measurements. First, a fixed-focus camera was used (without the EVLCL) and the position of the object was changed between four positions. Images were recorded and the corresponding MTF data were generated for each position. The results obtained are shown in Figure 6.25. For a simpler presentation, we use "optical power settings" in diopters corresponding to the four positions of the object. As can be seen (Figure 6.25, open squares), in the case of the "fixed-focus lens," the type of base lens and its distance from the image sensor provide the best focusing at approximately 67 cm, corresponding to ≈1.5 D. We obviously observe a drastic drop of the MTF for closer and further object distances. The next measurement was done by adjusting (using a micrometric stage) the position of the base lens (with respect to the image sensor) for each position of the object. This would correspond to the ideal case of AF (Figure 6.25, filled circles). Correspondingly, the MTF remained relatively high and did not degrade noticeably when the object distance was changed.

Finally, an EVLCL was added in front of the base lens and the OP of this lens was electrically changed (without mechanical adjustment of the base lens) for each position of the object. The corresponding results for the MTF, averaged across the field of view, show (Figure 6.25, open triangles) that, while there is some degradation when moving from infinity to approximately 11 cm

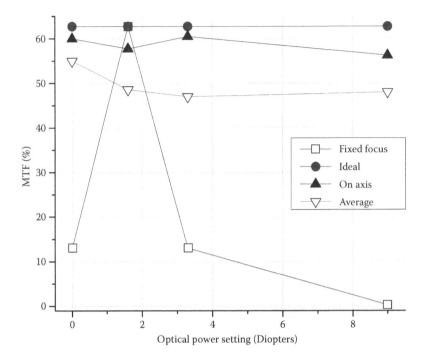

FIGURE 6.25 MTF measurements performed for different conditions. Circles: ideal autofocus; squares: fixed focus; open triangles: using EVLCL, averaged across the field of view; filled triangles: using EVLCL, on axes. (From LensVector. With permission.)

(corresponding to ≈9 D), the overall MTF remains quite acceptable for all object distances. Note also that the obtained on-axes performance (Figure 6.25, filled triangles) is significantly better (<3% of degradation compared to the base lens).

6.2.3.2 AFCT

As we have already mentioned, the transition times of EVLCLs between different OP levels have very interesting specificities. Indeed, Figure 6.20 shows that those transitions are usually smooth and without ringing effects. This feature (combined with the absence of the hysteresis and the orientation independence) allows the use of very simple and efficient driving algorithms to achieve extremely short AFCTs both for still pictures and for video regimes. Thus, LensVector has developed a proprietary AF algorithm (Figure 6.26), which leverages the above-mentioned advantages of EVLCLs. Commercially available, high-quality cameras with a similar format were selected for comparison to demonstrate the obtained results. The AFCTs were measured using the Sofica

"CamSpeed" application (average of 10 AF sequences with office lighting, approximately 400 lux). The EVLCL's performance was measured using a comparable camera development system. As can be seen from Figure 6.26, the performance of LensVector's AF system (see LV, Figure 6.26) is quite comparable with high-quality "smart" mobile-phone camera systems.

In addition, it is important to mention that the AFCT of LensVector's device has no noticeable dependence from the camera's frame rate, which is an important advantage compared to the currently available VCM solutions.

6.2.3.3 Veiling Glare

All real optical materials (including glass and polymer) scatter light (Bohren and Huffman 2004), which can degrade the camera's veiling glare (VG) performance by reducing the contrast between the "dark" and "bright" pixels of the image. In the majority of cases, these are optical nonuniformities (e.g., density fluctuations), which are at the origin of light scattering. However, in the case of NLC, such nonuniformities are mainly generated by thermal fluctuations of the director's orientation (Gennes and Prost 1995). The corresponding scattering cross section, σ, is directly proportional to the dielectric anisotropy at optical frequencies $(\Delta\varepsilon)^2$ and is inversely proportional to K; $\sigma \sim (\Delta\varepsilon)^2/K$. By an appropriate choice of NLC material and its thickness d, the scattering contribution may be noticeably reduced. This is why a lens-specific FoM must be developed to also take into account the achievable values of OP. In addition, higher driving voltages may

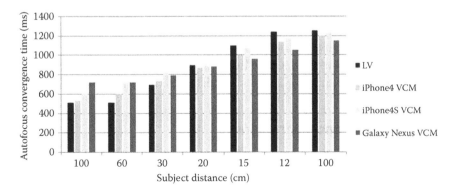

FIGURE 6.26 Autofocus convergence-time measurements for various mobile-phone cameras identified on the right. (From LensVector. With permission.)

also noticeably reduce the scatter. Thus, <1% of VG degradation was demonstrated, which, according to many experts, is a quite acceptable performance for the majority of mobile cameras.

6.3 Future Potential Developments and Applications

EVLCL components may provide more advanced functionalities than a simple lens. One possible way of creating such functionalities is the segmentation of control electrodes (Ye et al. 2006). This allows the generation of noncentrosymmetric phase-front profiles. Thus, following this avenue, the "adaptive optics" capacity of the EVLCL was demonstrated (Galstian et al. 2010), which allows smart wave-front manipulation by dynamically generating forms such as a prism, coma, and astigmatism.

In the particular case of the prismlike wave form (which is obtained when the NLC molecules are aligned parallel to the substrate in one corner and are aligned perpendicular to the substrate in the opposite corner of the lens, thereby forming a "molecular prism"), the NLC may be used to enable OIS functionality. In fact, the OIS may be done by the same element, which also provides AF functionality, making such a device even more attractive from the point of view of cost, power consumption, and size (such components may be produced with record small lateral sizes: 4.5×4.5 mm^2 compared to VCM-based 11×11 mm^2 OIS systems).

The aforementioned applications of EVLCLs are just examples. There may be various other applications provided that the CA of the camera is not very large (since the OP of the lens decreases as the square of the CA). Among others, the Blu-Ray disc system, ophthalmology (Fowler and Pateras 1990) (including contact and intraocular lenses), and endoscopic (microimaging, Raman probes, etc.) applications are well adapted for such lenses. Finally, microlens arrays have been studied in view of their applications in polarization-free LCDs (Lee et al. 2013) and in switchable 2D to 3D eye-wearing, glass-free television (Krijn et al. 2008).

6.4 Summary and Conclusions

I hope that the information provided in this chapter about EVLCLs would encourage people, students, and engineers who are curious about the technology to take a close look at EVLCLs and to make the right choice for their application. As with any other technology, EVLCLs have advantages and drawbacks.

Their strongest value proposition is their low cost at a rather good optical performance (their VG, AFCT, and OP performance are quite comparable with other technologies). Their main challenge, which is typical for all GRIN lenstype solutions, is the preservation of MTF when changing the level of OP. Indeed, the addition of a variable element on the top of the "perfectly" designed base lens will always introduce some degree of MTF degradation (even the mechanical movement of the base lens does so). However, the smart wave-front manipulation, enabled by LensVector's EVLCLs, allows significant mitigation of this problem. EVLCLs have other "secondary" advantages (such as silent operation and low power consumption) and drawbacks (such as VG and low-temperature AFCT). As has already been mentioned, the scatter and low-temperature issues of EVLCLs are resolved by, respectively, the appropriate choice of material parameters and higher working temperatures (in fact, the internal temperature of mobile-camera modules is already rather high, at the order of 50°C, due to the high degree of miniaturization and integration of its components).

Acknowledgments

This chapter was made possible thanks to the collaboration of many of my students, postdocs, and colleagues, in the research laboratories of both Université Laval and LensVector. I have to particularly emphasize the names of K. Asatryan, V. Presnyakov, A. Tork, A. Zohrabyan, and A. Bagramyan. The early work of our group on EVLCLs started at Université Laval thanks to the financial support of NSERC (Canada), FCAR (Quebec), and Sovar (Quebec) and was then further developed up to the first commercially available consumer electronic product by the R&D and engineering teams of LensVector. I am thus deeply thankful to all aforementioned colleagues and institutions for their extraordinary collaboration and help.

References

Asatryan, K., Presnyakov, V., Tork, A., Zohrabyan, A., Bagramyan, A., and Galstian, T. (2010), Optical lens with electrically variable focus using an optically hidden dielectric structure, *Optics Express 18*(13), 13981–13992.

Blinov, L. M. and Chigrinov, V. G. (1994), *Electrooptic Effects in Liquid Crystal Materials*, Springer, Berlin.

Bohren, C. F. and Huffman, D. R. (2004), *Absorption and Scattering of Light by Small Particles*, Wiley VCH, New York, Volume 1.

Boss, P., Bryant, D., Shi, L., and Wall, B. (2011), Tunable electro-optical liquid crystal lenses and methods for forming the lenses, U.S. Patent Application 2011/0025955 A1.

Bricot, C., Hareng, M., and Spitz, E. (1977), U.S. Patent 4,037,929.

Commander, L. G., Day, S. E., and Selviah, D. R. (2000), Variable focal length microlenses, *Optics Communications 177*, 157–170.

Doyle, O. and Galstian, T. (2009), Photo induced gradient index lenses with W profile, *Optics Express 17*(7), 4970–4975.

Fan, Y.-H., Ren, H., and Wu, S.-T. (2003), Switchable Fresnel lens using polymer-stabilized liquid crystals, *Optics Express 11*(23), 3080.

Fowler, C. W. and Pateras, E. S. (1990), Liquid crystal lens review, *Ophthalmic and Physiological Optics 10*, 186–194.

Galstian, T. (2010), Liquid crystal lens using surface programming, PCT, WO 2010/006420 A1.

Galstian, T., Asatryan, K., Tork, A., Presniakov, V., Zohrabyan, A., and Babramyan, A. (2012a), Optically hidden electromagnetic source for generation of spatially non uniform magnetic field and tunable devices made thereof, U.S. Patent 8,184,218.

Galstian, T., Zohrabyan, A., Asatryan, K., Tork, A., Presniakov, V., and Bagramyan, A. (2012b), Thermal liquid crystal optical device, U.S. Patent 8,184,217.

Galstian, T., Presniakov, V., Asatryan, K., and Zohrabyan, A. (2010), Image stabilization and shifting in a liquid crystal lens, U.S. Patent Application 20120257131.

Galstian, T., Presniakov, V., Asatryan, K., Tork, A., Zohrabyan, A., and Bagramyan, A. (2011), Electro-optical devices using dynamic reconfiguration of effective electrode structures, U.S. Patent, 8,033,054 B2.

Gennes, P. G. and Prost, J. (1995), *The Physics of Liquid Crystals*, Oxford University Press, Cambridge, 2nd edn.

Goodman, J. W. (1968), *Introduction to Fourier Optics*, McGraw-Hill, New York.

Hashimoto, N. (2009), Liquid crystal optical element and method for manufacturing thereof, U.S. Patent 7,619,713 B2.

Hecht, E. (2001), *Optics*, Addison-Wesley, Reading, MA, 4th edn.

Hensler, J. R. (1975), Method of producing a refractive index gradient in glass, U.S. Patent 3,873,408.

Kato, Y. and Kawada, T. (2007), Automatic focusing apparatus, U.S. Patent Application, U.S. 2007/0268417 A1.

Khoo, C. and Wu, S. T. (1993), *Optics and Nonlinear Optics of Liquid Crystals*, World Scientific, Singapore, Chapter 2.

Knittel, J., Richter, H., Hain, M., Somalingam S., and Tschudi, T. (2005), Liquid crystal lens for spherical aberration compensation in a blu-ray disc system, *IEE Proceedings Science, Measurement and Technology 152*(1).

Krijn, M. P. C. M., Zwart, S. T., Boer, D. K. G., Willemsen, O. H., and Sluijter, M. (2008), 2-D/3-D displays based on switchable lenticulars, *Journal of the Society for Information Display 16*(8), 847–855.

Lee, C.-T., Li, Y., Lin, H.-Y., and Wu, S.-T. (2011), Design of polarization-insensitive multi-electrode GRIN lens with a blue-phase liquid crystal, *Optics Express 19*(18), 17402–17407.

Lee, Y.-J., Baek, J.-H., Kim, Y., Heo, J. U., Moon, Y.-K., Gwag, J. S., Yu, C.-J., and Kim, J.-H. (2013), Polarizer-free liquid crystal display with electrically switchable microlens array, *Optics Express 21*(1), 129–134.

Merck Product Specifications. http://www.merck-chemicals.com/lcd-emerging-technologies.

Moore, D. T. (1977), Design of single element gradient-index collimator, *Journal of Optical Society of America 67,* 1137-1143.

Moore, D. T. (1980), Gradient-index optics: A review, *Applied Optics 19*(7), 1035–1038.

Naumov, A. F., Loktev, M. Yu., Guralnik, I. R., and Vdovin, G. (1998), Liquid-crystal adaptive lenses with modal control, *Optical Letters 23,* 992–994.

Naumov, A. F., Love, G. D., Loktev, M. Yu., and Vladimirov, F. L. (1999), Control optimization of spherical modal liquid crystal lenses, *Optics Express 4*(9), 344–352.

Nose, T., Masuda, S., Sato, S., Li, J., Chien, L.-C., and Boss, P. J. (1997), Effects of low polymer content in a liquid-crystal microlens, *Optics Letters 22*(6), 351–353.

Presnyakov, V. V., Asatryan, K. E., Galstian, T., and Tork, A. (2002), Tunable polymer-stabilized liquid crystal microlens, *Optics Express 10*(17), 865–870.

Ren H. and Wu, S.-T. (2003), Tunable electronic lens using a gradient polymer network liquid crystal, *Applied Physics Letters 82*(1), 22–24.

Ren, H., Fan, Y.-H., Gauza, S., and Wu, S.-T. (2004), Tunable-focus flat liquid crystal spherical lens, *Applied Physics Letters 84*(23), 4789–4791.

Reznikov, M., Reznikov, Yu., Slyusarenko, K., Varshal, J., and Manevich, M. (2012), Adaptive properties of a liquid crystal cell with a microlens-profiled aligning surface, *Journal of Applied Physics 111,* 103118.

Riza, N. A. and Dejule, M. C. (1994), Three-terminal adaptive nematic liquid-crystal lens device, *Optics Letters 19*(14), 1013–1015.

Sato, S. (1979), Liquid-crystal lens-cells with variable focal length, *Japanese Journal of Applied Physics 18*(9), 1679–1684.

Sato, S. (1999), Applications of liquid crystals to variable-focusing lenses, *Optical Review 6*(6), 471–485.

Sato, S., Sugiyama A., and Sato, R. (1985), Variable-focus liquid-crystal Fresnel lens, *Japanese Journal of Applied Physics 24*, L626–L628.

Takatoh, K., Hasegawa, M., Koden, M., Itoh, N., Hasegawa, R., and Sakamoto, M. (2005), *Alignment Technologies and Applications of Liquid Crystal Devices*, Taylor & Francis, London.

Tseng, M.-C., Fan, F., Lee, C.-Y., Murauski, A., Chigrinov, V., and Kwok, H.-S. (2011), Tunable lens by spatially varying liquid crystal pretilt angles, *Journal of Applied Physics 109*, 083109.

Valley, P., Mathine, D. L., Dodge, M. R., Schwiegerling, J., Peyman, G., and Peyghambarian, N. (2010), Tunable-focus flat liquid-crystal diffractive lens, *Optics Letters 35*(3), 336–338.

Vdovin, G. V., Guralnik, I. R., Kotova, S. P., Loktev, M. Y., and Naumov, A. F. (1999), Liquid-crystal lenses with a controlled focal length. I: Theory, *Quantum Electron 29*, 256–260.

Wang, B., Ye, M., Honma, M., Nose, T., and Sato, S. (2002), Liquid crystal lens with spherical electrode, *Japanese Journal of Applied Physics 41*(11A), 1232–1233.

Wang, B. Ye, M., and Sato, S. (2004a), Numerical study of a lens-shaped liquid crystal cell, *Molecular Crystal and Liquid Crystal 413*, 423[2559]–433[2569].

Wang, B., Ye, M., and Sato, S. (2004b), Lens of electrically controllable focal length made by a glass lens and liquid-crystal layers, *Applied Optics 43*(17), 3420–3425.

Wu, S.-T., Lackner, A. M., and Efron, U. (1987), Optimal operation temperature of liquid crystal modulators, *Applied Optics 26*(16), 3441–3445.

Ye M. and Sato, S. (2002), Optical properties of liquid crystal lens of any size,' *Japanese Journal of Applied Physics 41*(5B), 571–573.

Ye, M., Wang, B., and Sato, S. (2004), Liquid-crystal lens with a focal length that is variable in a wide range, *Applied Optics 43*(35), 6407.

Ye, M., Wang B., and Sato, S. (2006), Study of liquid crystal lens with focus movable in focal plane by wave front analysis, *Japanese Journal of Applied Physics 45*(8A), 6320–6322.

Ye, M., Wang, B., Uchida, M., Yanase, S., Takahashi, S., Yamaguchi, M., and Sato, S. (2010), Low-voltage-driving liquid crystal lens, *Japanese Journal of Applied Physics 49*, 100204.

Yeh, P. and Gu, C. (1999), *Optics of Liquid Crystal Displays*, Wiley, Hoboken, NJ.

CHAPTER 7

Optical Image Stabilization for Miniature Cameras

Eric Simon

Contents

7.1 Introduction

Miniature camera modules have become a key subsystem in the mobile-phone and miniature camera handset. The trend in camera phones has been to increase the pixel count, while maintaining small sensor formats for overall size and cost constraints. Pixel shrinking has led to degraded light sensitivity thus requiring increased exposure time that, in turn, impacts on the image quality due to hand-shake blur.* Electronic and software hand-shake blur

---

* Electronic image stabilization (EIS) can perform an efficient stabilization of video, compensating for the average image translation due to hand shake from one frame to the next (Kinugasa et al., 1990). However, EIS fails to correct the blur during the pixel integration time of a picture. It can only be corrected with an optical correction during the picture capture.

reduction are not always successful in restoring good image quality. Several groups have pointed out this major problem in mobile imaging, which can only be resolved by an optical image stabilization (OIS) system (Golik and Wueller 2007; Xiao et al. 2005, 2006, 2007, 2009; Shimohate 2002; Chiu et al. 2008) (see Figure 7.1).

The design of an efficient OIS system is highly dependent on the characteristic of the hand shake. Therefore, we will start this chapter with an example of an experimental hand-shake characterization performed in the context of smartphone cameras. The image blur generated by the photographer's hand shake during the exposure time of the image is very variable from one person to another, and for the same person, from one picture to another. We present a model of the average hand shake of a smartphone obtained from a statistical analysis of a large number of experimental recordings. In the following section, we analyze the contribution of the hand shake to the image blur and we give an estimation of the consecutive image blur in terms of number of pixels as a function of the exposure time and of the image sensor resolution. We present the intrinsic optical noise limits in the context of smartphone cameras. This theoretical study highlights the minimal picture integration time that is required for different shooting conditions in order to obtain pictures with an acceptable noise level and it points out when OIS is required. In the context of miniature smartphone cameras, this result can be compared to the photographer's rule of thumb giving the maximal exposure time to obtain sharp pictures with a 24 × 36 mm film format. We describe some OIS technologies for smartphones and a theoretical model simulating the blur reduction efficiency as a function of the OIS optical actuator characteristics and of the picture integration time is given in the last section.

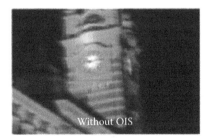

FIGURE 7.1 Example of a hand-shake blur reduction performed with an optical image stabilization.

7.2 Hand-Shake Characterization

Designing and simulating the performances of an OIS system require knowledge of the hand-shake characteristics: the amplitude, speed, and decomposition over the different translation and rotation components. We present a synthetic view of a hand-shake experimental characterization and modeling based on existing data and on test campaigns performed in our laboratory. Due to the physiological tremor of muscles (Golik and Wueller 2007), a human being cannot hold a camera still. The hand always moves slightly and these movements are responsible for the hand-shake blur on pictures. The hand shake can be decomposed into six motion components: three translation components δX_{hs}, δY_{hs}, and δZ_{hs} along the Ox, Oy, and Oz axes and three tilts around the Ox, Oy, and Oz axes: the pitch $\delta\Theta x_{hs}$, the yaw $\delta\Theta y_{hs}$, and the roll $\delta\Theta z_{hs}$ (see Figure 7.2).

We have estimated those six different motions in a test campaign with a panel of 16 healthy people aged between 20 and 45 years with two recordings per person. Each person was simply asked to hold the mobile phone as if he or she were taking a picture except that the button press action was not performed. The 130 g mass of the instrumented mobile phone was representative of the mass of an average smartphone. For the hand-shake rotation measurement, a collimated laser beam emitting a light spot every 25 ms was set on the mobile phone and the movement of the laser spot on a wall at a distance of 5 m was recorded on a picture taken with a 400 ms integration time (see Figure 7.3). For the hand-shake translation measurement, a noncollimated laser diode

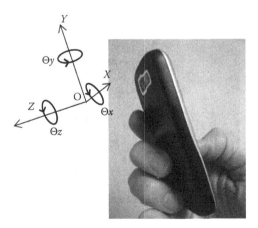

FIGURE 7.2 Hand-shake translation and tilt components.

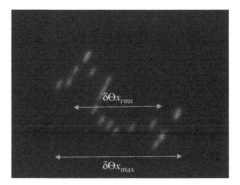

FIGURE 7.3 A 400 ms recording of a laser spot flashing every 25 ms on a wall at a far distance.

emitting a punctual light spot every 25 ms was set on the mobile phone and the movement of the laser diode spot was recorded on a 400 ms integration picture of the mobile phone held by the photographer (see Figure 7.4).

From the 32 picture recordings, we have performed a statistical evaluation of the mobile-phone tilt and translation during an integration time, T_{pct}, of 400 ms. We have considered the total amplitude from the initial position and the root mean square (RMS) amplitude for the three tilt components and the three translation components. Those respective expressions are given in Equations 7.1 and 7.2 for the example of the pitch tilt component $\delta\Theta x_{\max}$ and $\delta\Theta x_{rms}$.

$$\delta\Theta x_{\max}\left(T_{Pct}\right) = \max_{t:0\to T_{pct}}\left|\delta\Theta x_{hs}\left(t\right)\right|, \tag{7.1}$$

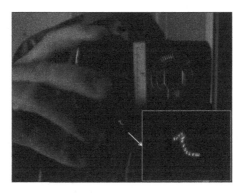

FIGURE 7.4 A 400 ms recording of a laser spot flashing every 25 ms on a handheld mobile phone for translation measurements.

$$\delta\Theta x_{rms}\left(T_{Pct}\right) = 2 \cdot \sqrt{\frac{1}{T_{pct}} \int_{t=0}^{t=T_{pct}} \left(\delta\Theta x\left(t\right) - \overline{\delta\Theta x_{T_{pct}}}\right)^2 dt} \quad (7.2)$$

where $\delta\Theta x(t)$ is the instantaneous pitch tilt of the camera. The total amplitude from the initial position gives an estimation of the compensation amplitude required by an OIS actuator to perform an efficient correction. The RMS amplitude, defined from the average position $\overline{\delta\Theta x_{T_{pct}}}$ during the picture integration time, gives a representative estimation of the RMS blur spot diameter on the image. The post-treatment of the 32 recordings has not shown any statistical differences between the translation amplitude along x or y in regard to the experimental dispersion. We do not observe any statistical differences between the amplitude of the three hand-shake tilt components around x, y, and z. For that reason, we have assumed, respectively, that the translations δX_{hs} and δY_{hs} and the tilts $\delta\Theta x_{hs}$, $\delta\Theta y_{hs}$, and $\delta\Theta z_{hs}$ have a statistically equivalent amplitude.

Table 7.1 shows a statistical analysis of a hand-shake tilt component and a hand-shake translation component during 400 ms based on 64 recordings with 16 people.

Figures 7.5 and 7.6, respectively, give an overview of the distribution of the tilt amplitude from its origin and the RMS tilt amplitude. A factor of 10 is observed between the smallest and the largest hand-shake tilt amplitude. It has been shown that the photographer's training (Xiao et al. 2006) or both hands holding the camera (Xiao et al. 2007) can statistically reduce the amount

TABLE 7.1 Statistical Analysis of the Hand-Shake Tilt and the Hand-Shake Translation

	Average Value over 64 Recordings	Standard Deviation σ over 64 Recordings
Tilt around (Ox) during 400 ms $\delta\Theta x_{max}$	0.32°	0.2°
RMS tilt amplitude around (Ox) during 400 ms $\delta\Theta x_{rms}$[a]	0.22°	0.13°
Translation along (Ox) during 400 ms δX_{max}	1.7 mm	1 mm
RMS translation along (Ox) during 400 ms δX_{rms}	1.1 mm	0.5 mm

[a] We can see empirically that $\Theta max_{400\,ms} \approx 3/2 \cdot \Theta_{rms-400\,ms}$.

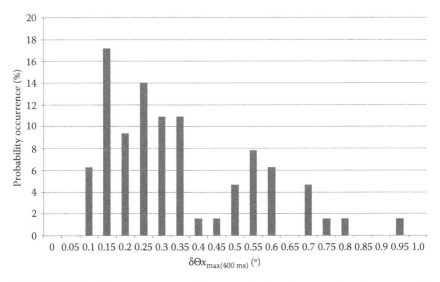

FIGURE 7.5 Probability occurrence of a given tilt from the origin (pitch, yaw, or roll component) during a 400 ms recording of a person holding a smartphone.

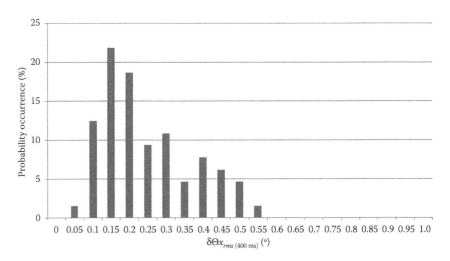

FIGURE 7.6 Probability occurrence of a given RMS tilt diameter (pitch, yaw, or roll component) during a 400 ms recording of a person holding a smartphone.

of average picture blur from 20% to 50%, but the dispersion of the blur amplitude from one picture to another still remains large. Although the dispersion of the blur amplitude is large, we can see from Figure 7.5 that no tilt hand-shake amplitude larger than 1° around one axis was observed during 400 ms and we can see also that in 90% of the picture captures, the tilt amplitude was below 0.6°. Assuming that 400 ms is an extreme integration time for picture captures and that our panel of people is well representative of casual photographers, we can consider that the blur of most pictures can be corrected with an OIS system that can compensate by at least a ±0.6° tilt amplitude around the Ox and Oy axes. The influence of hand-shake translation is discussed in the next section.

Simulating OIS performances requires an extension of previous results for different exposure times. We propose to use the following simple model to describe the RMS amplitude of the tilt as a function of the exposure time T_{pct}:

$$\delta\Theta x_{rms}\left(T_{Pct}\right) = \alpha_{\theta\_x} \cdot \left(T_{Pct}\right)^{b} \qquad (7.3)$$

The term α_{θ_x} is an amplitude coefficient and the term b is between the extreme values $b = 0.5$ (corresponding to a white noise tilt velocity of a random-walk tilt) and $b = 1$ (corresponding to a straight-line-walk pattern). Several experimental studies show that the hand-shake tilt with a mobile-phone camera is in between these two models. We found that a value of $b = 0.62$, which is slightly higher than the 0.56 value proposed by Xiao et al. (2006) for a digital still camera (DSC), gives a good agreement with our experimental data in the 50–400 ms range. From this statement and from the data in Table 7.1, a realistic model of the RMS tilt amplitude around (Ox) as a function of the exposure time is given as follows:

$$\delta\Theta x_{rms} \approx 0.37 \cdot \left(T_{Pct}\right)^{0.62} \qquad (7.4)$$

The exposure time, T_{pct}, is expressed in units of seconds and $\delta\Theta x_{rms}$ is measured in degrees. The yaw $\delta\Theta y_{rms}$ and the roll $\delta\Theta z_{rms}$ can be similarly described.

The power spectral density of the hand shake is also helpful data for an OIS performance simulation. The random-walk tilt model with $b = 0.5$ corresponds to a monolateral power spectral density $S_{\Theta x_{hs}}(\nu)$ inversely proportional to the square of the frequency:

$$S_{\delta\Theta x_{hs}}\left(\nu\right) = \frac{a_{hs}}{\nu^{2}} \qquad (7.5)$$

where a_{hs} is an amplitude coefficient in units of Hz (°)². Compared to this theoretical model, our experimental fast Fourier transforms of 1 s gyroscope recordings with a handheld mobile phone show a larger contribution of the low frequency noise and we find the power spectral density of hand-shake tilt fluctuations to be much better described with the following model:

$$S_{\delta\Theta x_{hs}}(\nu) \approx \frac{a_{\exp}}{\nu^{2.25}} \tag{7.6}$$

where $a_{\exp}$ is an amplitude coefficient in units of Hz1.25 (°)².

7.3 Contribution of Hand Shake to Image Blur

The motion of the camera during a picture exposure time creates an image blur, which can be expressed from the focal length, f, of the camera lens and from the distance, D_{obj}, between the camera and the object in the focus plane (Figure 7.7).

For given hand-shake translations δX_{hs}, δY_{hs}, and δZ_{hs} and given hand-shake tilts $\delta\Theta x_{hs}$, $\delta\Theta y_{hs}$, and $\delta\Theta z_{hs}$ of the camera during a picture exposure time, the displacement (δx, δy) of the image of a punctual object at a distance D_{obj} from the camera can be expressed as in Equation 7.7:

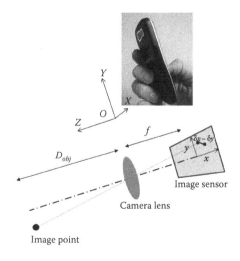

FIGURE 7.7 Schematic of a handheld picture capture with a mobile phone.

$$\delta x = f \cdot \tfrac{\pi}{180} \cdot \delta \Theta y_{hs} + \frac{f}{D_{obj}} \cdot \delta X_{hs} + y \cdot \tfrac{\pi}{180} \cdot \delta \Theta z_{hs}$$

$$\delta y = -f \cdot \tfrac{\pi}{180} \cdot \delta \Theta x_{hs} + \frac{f}{D_{obj}} \cdot \delta Y_{hs} - x \cdot \tfrac{\pi}{180} \cdot \delta \Theta z_{hs}$$

(7.7)

where (x,y) are the Cartesian coordinates of the punctual image in the sensor plan. The tilts are measured in degrees.

The blur is composed of three components:

- The first components $f \cdot \tfrac{\pi}{180} \cdot \delta \Theta x_{hs}$ and $f \cdot \tfrac{\pi}{180} \cdot \delta \Theta y_{hs}$ are, respectively, induced by the hand-shake pitch and the hand-shake yaw of the camera. These are the major contributors of blur on most pictures. Their compensation can be performed with an optical tilt actuator. This necessitates a real-time measurement of the camera tilt around (Ox) and (Oy) during the picture capture with a dual-axis gyroscope.
- The second components, $f/D_{obj} \cdot \delta X_{hs}$ and $f/D_{obj} \cdot \delta Y_{hs}$, correspond to the blur induced by a hand-shake translation of the camera during the picture capture. This blur contribution is predominant for macrophotography (see Figure 7.8) but becomes negligible when the object is at a far distance from the camera. The correction of the blur requires a translation sensor such as an accelerometer. There is a critical object distance, D_{crit}, under which the image blur induced

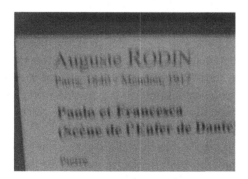

FIGURE 7.8 Example of an image blur on a picture taken at a distance of approximately 10 cm with a DSC. The OIS of this camera compensates for hand-shake tilts whereas the blur is mostly induced by a vertical translation of the camera, which is not detected by the gyroscope.

by the translation has the same order of magnitude as the image blur induced by the tilt:

$$D_{crit} = \frac{\delta X_{hs}}{\frac{\pi}{180} \delta \Theta y_{hs}}$$ (7.8)

From a statistical analysis of the RMS values of the hand-shake translation, δX_{hs}, and the hand-shake tilt, $\delta \Theta x_{hs}$, during 400 ms, presented in Table 7.1, we estimate that the D_{crit} is around 30 cm. For pictures of objects at a distance of more than 1 m, the image blur is mainly due to the tilt of the camera and an efficient OIS system can be designed with a gyroscope sensor only. This case corresponds to the majority of pictures. For pictures of objects at a distance smaller than 10 cm, an efficient OIS for hand-shake correction requires an accelerometer sensor to correct the blur induced by hand-shake translation movements of the camera. For pictures of objects at a distance between 10 cm and 1 m, an efficient OIS requires an estimation of the object distance and a measurement of both tilt and translation during the image capture.

■ The third terms, $y \cdot \frac{\pi}{180} \cdot \delta \Theta z_{hs}$ and $x \cdot \frac{\pi}{180} \cdot \delta \Theta z_{hs}$, correspond to the blur induced by the roll: the rotation of the camera around the optical axis. With most OIS designs, these terms cannot be compensated for. The blur contribution depends on the position of the object in the field. In the center of the image, no blur is generated from the roll movement. At the border of the image, the blur of the roll movement has a maximum value depending on the field of view (FOV). If we consider a 63° diagonal image FOV,* the blur contribution of the roll $x \cdot \frac{\pi}{180} \cdot \delta \Theta z_{hs}$ at the border of the image is approximately $\frac{f}{2} \cdot \frac{\pi}{180} \cdot \delta \Theta z_{hs}$. Since the three tilt contributions statistically have the same order of magnitude, we can see from Equation 7.7 that in this worst case, the blur due to roll will remain twice as small as the pitch or yaw term.

For pictures of objects at a distance of more than 1 m, we can assume that the blur in the center of the image is mostly due to pitch and yaw hand-shake tilts. With a standard 4/3 image

---

* A 63° diagonal field of view corresponds to an equivalent 35 mm focal length with 24 × 36 standard film format. It is the most usual field of view for smart phone cameras.

format, the magnitude of the blur δx given in Equation 7.7 can be expressed in units of number of pixels as a function of the total number of megapixels of the image $nMpix$ and of the image diagonal FOV as follows:

$$n_{pix}x = \frac{\frac{\pi}{180} \cdot \delta\Theta x_{hs} \cdot \frac{5}{4\cdot\sqrt{3}} \cdot \sqrt{nMpix}}{\tan\left(\frac{1}{2} \cdot FOV\right)} \tag{7.9}$$

The expression for the blur along δy is similar. Assuming a standard 4/3 image format with a 63° diagonal object FOV and a blur amplitude as described in Equation 7.4, from Equation 7.9 we can simply express an evaluation of the number of pixels blur along the x- or y-direction as a function of the picture integration time and of the number of megapixels of the image:

$$n_{pix}x \approx 7.9 \cdot T_{pct}^{0.62} \cdot \sqrt{nMpix} \tag{7.10}$$

7.4 Intrinsic Limit of Smartphone Cameras for Indoor Pictures

If we compare smartphone cameras to other types of cameras, there is one crucial difference: the thickness. The average thickness of a smartphone is usually smaller than 1 cm.* Independent from the design choices and the possible technological advances, this thickness constraint imposes fundamental limits on the minimum exposure time required to take good pictures. This element makes the implementation of an OIS in smartphones a key advantage.

Assuming a 1 cm smartphone thickness hypothesis, if we allocate a budget of about 4 mm for the sensor circuit, the housing, and the optical windows, an approximate thickness of 6 mm remains for the total size of the camera lens, corresponding approximately to a maximal 5 mm focal length with conventional lens designs. With this focal length, a standard 63° diagonal FOV gives a 6.1 mm image diagonal corresponding to the 1/3-inch standard sensor format. Using a bigger sensor with a 5 mm focal length lens is possible by increasing the FOV, but we can consider that a half-inch sensor, corresponding to a 75° diagonal FOV, equivalent

---

* A significant change of this constraint would require a change in today's smartphone standards or major technical progress in miniature telescopic lens designs.

TABLE 7.2 Pixel Sizes of Different 4/3 Format Image Sensors

Megapixels	Pixel Size (μm) According to Standard 4/3 Sensor Formats (μm)			
	1/4-inch	1/3.2-inch	1/2.5-inch	24 × 36 mm Reflex Sensor
3.2	1.75	2.2	2.8	
5.1	1.4	1.75	2.2	
8.1	1.1	1.4	1.75	
12.6		1.1	1.4	
21			1.1	9

to 28 mm in a 24 × 36 film standard, would reasonably be the maximum sensor size with a 1 cm thick smartphone.* With the maximum size of the sensor fixed, increasing the number of pixels necessarily reduces the pixel size and for a given picture integration time, the amount of photons received per pixel is also reduced proportionally to the inverse of the active pixel surface. Table 7.2 shows the pixel size corresponding to different image resolutions with different sensor sizes. We have selected three of the largest standard sizes of smartphone sensors—1/4-inch, 1/3.2-inch, and 1/2.5-inch—and for comparison, the 24 × 36 mm² sensor size of a reflex camera.[†]

We have considered pixel sizes down to 1.1 μm. With such a small pixel size and a f/2.4 optical aperture, the true image resolution is limited not only by the number of pixels but also by the fundamental optical diffraction limit.[‡] Due to the optical diffraction limit, for a given sensor size, increasing the number of pixels by reducing the pixel size smaller than 1.1 μm will not improve the image resolution.

Table 7.2 outlines the major difference between a reflex camera and a high-resolution smartphone: there is at least a factor of 40 on the pixel surface reduction. For a given scene, a given image resolution, a given optical aperture, and a given picture exposure time, the result will be at least the same factor of 40 between the number of photons detected on a pixel of a reflex camera and those

---

* The very wide field of a 28 mm focal length or smaller probably no longer corresponds to the expectations of casual photographers.
[†] The indicative 4/3 image formats for 1/4-inch, 1/3.2-inch, and 1/2.5-inch sensor formats are, respectively, 3.62 × 2.71 mm², 4.55 × 3.41 mm², and 5.78 × 4.33 mm².
[‡] For a perfect optical lens with a f/2.4 optical aperture, the half-width intensity of a diffraction spot at a wavelength of 550 nm is 1.2 μm.

on a pixel of a smartphone.* If the number of photons detected on each pixel becomes too small, the pixel signal-to-noise ratio will be degraded and the image will be noisier and less sharp due to the electronic read noise and, more fundamentally, to the quantum limit of photonic noise. A possible criterion (Xiao et al. 2005) for good-quality pictures is to detect a minimum of 1000 photons on an average brightness pixel.[†] In Figure 7.9, we have estimated the exposure time required to detect 1000 photons on a pixel for different scenes conditions as a function of the pixel size. We assume that no flash is used and we consider an optical aperture of f/2.8 for the lens, a 35% quantum efficiency of the sensor, and an 18% average scene reflectance.

The smaller the pixel size is, the longer the required exposure time and the larger the hand-shake blur would be. With a 24 × 36 film reflex camera, a photographer's rule of thumb gives an estimation of the maximal exposure time to avoid blurry pictures.[‡] A direct transposition of this rule to digital imaging is not possible for two reasons. First, hand shake is significantly smaller when holding a reflex camera than when holding a smartphone. Second, the image quality degradation by the hand-shake blur is a relative effect that must be compared to the maximal image resolution, significantly different with a 24 × 36 film and with a pixelated sensor. With a digital camera, the most relevant criterion is to compare the blur diameter to the size of a pixel, which is the intrinsic limit of the image resolution. With a color digital sensor,[§] let us take the simple hypothesis that a hand-shake blur with a RMS diameter smaller than one pixel has negligible impact on the image quality and let us consider a criterion of two pixels as a maximum acceptable RMS hand-shake blur diameter. The exposure time causing an average RMS blur diameter of n pixels in the case of

---

* With most image sensors, the difference is increased by the better filling factor achieved with large pixels.
[†] For a detection of 1000 photons, the read noise is usually smaller than the photonic noise and the pixel detection signal to noise is around 32 corresponding to the ultimate photonic noise limit.
[‡] The standard photographer's rule of thumb is "while shooting handheld, the shutter speed should never be lower than the focal length of the lens you're shooting with." This would correspond to a maximum 1/35 s, or 30 ms integration time with the 35 mm equivalent focal length reflex camera. Xiao et al. (2007) has proposed an extension of this rule to the larger hand shake observed with mobile phones. This study estimates a duration of 7.5 ms while holding the camera with one hand or 9.5 ms while holding it with two hands.
[§] We can estimate that a color sensor with a Bayer filter will have the same resolution as a monochrome sensor with twice as big pixels (Yotam et al. 2007; Simon et al. 2010a).

a 4/3 image format with a 63° diagonal FOV can be expressed from Equation 7.10:

$$T_{pct} \approx \left(\frac{n_{pix} x}{7.9 \cdot \sqrt{nMpix}} \right)^{1.61} \tag{7.11}$$

Table 7.3 gives numerical estimations of the exposure time corresponding to an average RMS diameter of, respectively, a one- or two-pixel RMS hand-shake blur diameter for different image sensor resolutions.

Based on those data and assuming a two-pixel RMS blur diameter to be acceptable with a color picture, we can see from Figure 7.9 that with natural lighting:

- Hand-shake blur is never a problem with sunny outdoor pictures
- Hand-shake blur is always a problem with night pictures
- Indoor pictures that are usually not limited by hand shake with the large pixels of a reflex camera will most probably become blurry with the small pixels of a smartphone camera

A reduction of the exposure time to perform the 1000 photon limit is theoretically possible with a larger optical aperture. But an aperture increase from f/2.8 to f/2 would represent a huge technical challenge for miniature lens designers and would bring a relatively modest factor of 2 on the exposure time reduction. This would not drastically change the previous statement. The use of the additional lighting of a flash lamp can improve the image sharpness but it is only efficient for relatively close distances,

TABLE 7.3   Estimation of the Average Exposure Time Corresponding to, Respectively, One or Two RMS-Pixel Blurs along the *x*- or *y*-Axis

Megapixels	Exposure Time Corresponding to An Average One-Pixel RMS Hand-Shake Blur Diameter (ms)	Exposure Time Corresponding to An Average Two-Pixel RMS Hand-Shake Blur Diameter (ms)
3.2	14	43
5.1	10	29
8.1	7	20
12.6	5	14
21	3	9

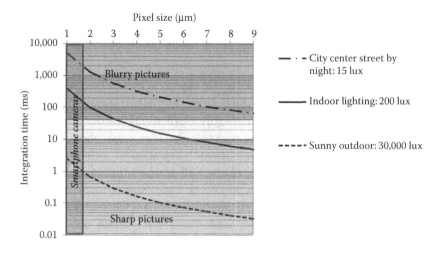

FIGURE 7.9 Integration time required to detect an average of 1000 photons on an average brightness pixel as a function of the pixel size (the area delimited by the vertical rectangle on the left side corresponds to smartphone cameras).

especially with the limited size and limited electrical power of a smartphone. When the flash is inefficient or when natural lighting is preferred, OIS is the only solution to recover a sharp image for indoor pictures with a mobile-phone camera.*

7.5 OIS Technologies

7.5.1 Introduction

All the available OIS systems are based on an actuator inducing a translation of the image on the image sensor. In a first approximation, a translation of the image on the sensor of (dx, dy) is equivalent to a tilt of the camera as described in Equation 7.7. With a 5 mm focal length, a ±80 µm image translation is equivalent to a ±1° tilt of the optical axis.

There are four main designs for OIS (Shimohata et al. 2002):

- Tilting the whole camera module
- Inducing an optical tilt with a variable prism mounted on the front of the camera lens

---

* When the optical power of the flash is insufficient to reduce the exposure time in order to make the hand-shake blur negligible, a combination of lighting and the OIS during the exposure time can restore a good image quality. Miniature LEDs are well adapted to this picture capture mode.

- Translating the image sensor relative to the camera lens
- Translating the camera lens or a part of the lens system relative to the image sensor

The last two solutions are the most commonly used in DSCs or reflex cameras. Due to the extreme miniaturization of the smartphone camera lens, it is highly challenging to directly transfer DSC and reflex camera technologies into smartphone cameras. In the following, we present three examples of technologies specifically developed to perform OIS in the miniature cameras of smartphones. These technologies have been shown to operate in prototypes and can all provide at least a tilt compensation of $\pm0.5°$. Similar and other interesting examples of optical actuators for OIS in miniature camera modules are presented by Kim et al. (2011), Song et al. (2010), Sachs et al. (2006), Kauhanen and Rouvinen (2006), and Yu and Liu (2007).

7.5.2 Lens Translation with Voice Coil Actuators

This technology consists in translating the camera lens along the x- and y-direction relative to the image sensor with voice coil actuators (Chiu et al. 2008). The lens is mounted on a holder which is maintained with springs along the x- and y-direction and the lens holder can be moved along yokes in the x- and y-direction with voice coil motors (a schematic is presented in Figure 7.10). The total thickness of the system can be as small as the lens thickness and the footprint size is around 15×15 mm^2.

The frequency response of the system can be approximately considered as a second-order resonant filter. The resonance frequency is imposed by the total mass of the moving part and by the stiffness of the equivalent spring. The dumping factor is determined

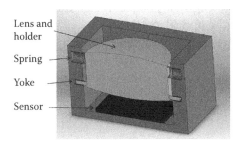

FIGURE 7.10 Schematic of an OIS mechanism based on a voice coil actuation of the lens.

by the friction force of the lens holder. The OIS can be efficient for a frequency below a fraction of the resonance frequency. Since the margin of optimization on the mass of the lens is limited, the stiffness of the spring is a key design element for the optimization of the maximal frequency response of the OIS. For given maximal values of the electromagnetic force and the electrical power consumption, a low stiffness spring enables a large amplitude correction at low frequencies, but it is inefficient for higher frequencies correction. On the contrary, a high stiffness spring is efficient at higher frequencies but it exhibits smaller amplitude corrections at lower frequencies. A higher stiffness also reduces the sensitivity to shocks and gravity.

The mechanical adjustment of the yoke and the mobile part is also a critical optomechanical element of the design. A tight adjustment will increase the friction and the hysteresis of the stroke. On the contrary, an excessive clearance between the yoke and the lens holder will induce lens tilt and image quality degradation.

7.5.3 Tilt of Whole Camera Module with Voice Coil Actuators

The principle is to tilt the whole camera module. This design can theoretically perform a large tilt correction with no optical degradation because the relative position of the camera lens and of the image sensor is fixed. It necessitates managing the electrical connection of the image sensor, which is moving relative to the printed circuit board (PCB) of the smartphone. Ahn et al. (2010) has proposed a miniature device adapted to the smartphone. The camera module includes a magnet on each side and is surrounded by four voice coils. The pitch and the yaw rotations of the camera module are induced by a differential command on a pair of voice coils (Figure 7.11).

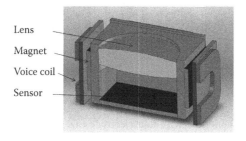

Lens

Magnet

Voice coil

Sensor

FIGURE 7.11   Schematic of a camera module tilted with voice coil actuators.

7.5.4 Liquid Lens Technology

A liquid lens is composed of two liquids with the same density: one is an electrically insulating liquid such as oil, and the other is an electrolyte (see Chapter 5, this volume). They have a refractive index difference to form an optical interface having an optical power depending on the curvature radius of the liquid interface. These two liquids rest on a hydrophobic and dielectric coating. When voltage is applied to the dielectric coating, the wettability of the liquids is modified and the shape of the liquid interface changes. This phenomenon is highly reversible with low hysteresis. When the same voltage is applied on the whole surface of the dielectric coating, the shape of the liquid interface remains both spherical and centered on the conical cavity symmetry axis. When a nonuniform voltage is applied along the dielectric coating, it induces a tilt of the liquid interface equivalent to an optical prism (Figure 7.12a). A nonuniform voltage is generated along

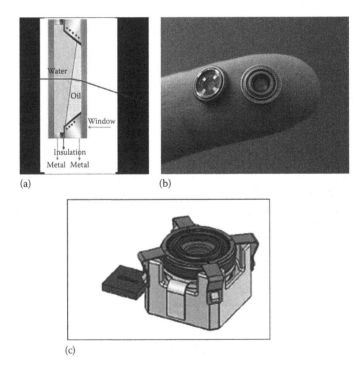

(a) (b)

(c)

FIGURE 7.12 OIS liquid lens with four electrodes: (a) schematic, (b) component, and (c) integration on a camera module.

TABLE 7.4 Example of the Performances of a Liquid Lens Component Dedicated to OIS for Miniature Camera Modules

Item	Value
Entrance pupil diameter	1.6 mm
Optical tilt range	±0.6°
Full range tilt response time @ 90%	30 ms
Focus range	−5 to 15 diopters

the cone with a design featuring at least four electrodes (see Figure 7.12b; Simon et al. 2010c).

The liquid lens is a modular component. It can be simply plugged on an existing camera module (Figure 7.12c) with an additional thickness of 1–2 mm in order to generate a focus control and a pitch and yaw tilt control of the optical axis (see Figure 7.12a). With no moving mechanical parts, this component is insensitive to mechanical shocks and it has low power consumption. Table 7.4 shows an example of the performance of a liquid lens dedicated to OIS. The experimental tilt transfer function can be interpolated in a good approximation with the following complex second-order filter model:

$$A_{LL}(v) = \frac{1}{1 + 2jm \cdot \dfrac{v}{v_0} - \left(\dfrac{v}{v_0}\right)^2} \tag{7.12}$$

where
  v is the frequency in units of hertz
  m is the damping factor
  v_0 is the resonance frequency in units of hertz.

7.6 OIS Performance Simulation

We present a theoretical estimation of the average hand-shake blur reduction performed by an OIS system (Simon et al. 2010b). It is based on hand-shake noise spectral density models and on the frequency response model of the optical tilt actuator. We consider only the pitch and the yaw hand-shake tilt components, assuming that the object distance is large enough to neglect the image blur due to translation hand shake. We do not consider the blur induced by the roll hand shake that cannot be compensated with the optical actuators previously described. As discussed in the previous section, it will bring additional blur in the border field of the image. For the pitch tilt component, the RMS amplitude of the angular hand-shake tilt generated during a picture integration time, T_{pct}, is given by Equation 7.2. The expression for the yaw hand-shake tilt component is similar. The average RMS value of the angular hand-shake pitch tilt generated during a picture integration time, T_{pct}, can also be expressed from the power spectral density of the hand-shake tilt with the following expression:

$$\delta\Theta x_{rms}\left(T_{pct}\right)=2\cdot\sqrt{\int_{v=0}^{v=\infty}S_{\delta\Theta x_{hs}}\left(v\right)\cdot\left|A_{T_{pct}}\left(v\right)\right|^{2}\cdot dv} \qquad (7.13)$$

where the term $A_{T_{pct}}\left(v\right)$ is the amplitude blur generated during T_{pct} by a unitary sine wave component at a frequency v averaged over a random initial phase.

$$\left|A_{T_{pct}}\left(v\right)\right|^{2}=1-\left(\frac{\sin\left(\pi\cdot v\cdot T_{pct}\right)}{\pi\cdot v\cdot T_{pct}}\right)^{2} \qquad (7.14)$$

The principle of an OIS system is to measure the instantaneous hand-shake tilt and to generate an opposite tilt with an optical tilt actuator, as shown in Figure 7.13.

If the maximal hand-shake tilt does not exceed the maximal optical tilt correction range of the actuator during the picture integration time, the correction loop response can be described with a linear model. According to the previous paragraph, a $\pm 0.6°$ range for pitch and yaw is necessary to efficiently correct the blur of most pictures.*

The power spectral density of the hand-shake tilt measurement can be expressed as follows:

$$S_{\delta\Theta x_{meas}}\left(v\right)=S_{\delta\Theta x_{gyro\_noise}}\left(v\right)+F_{gyro}\left(v\right)\cdot S_{\delta\Theta x_{hs}}\left(v\right) \qquad (7.15)$$

---

* If the maximal tilt from the origin is larger than the OIS tilt dynamic, the OIS efficiency is reduced but it can still provide a significant image quality improvement. The OIS works normally during the part of the picture integration time when the tilt is smaller than the maximum tilt and the OIS provides a saturated correction during the part of the picture integration time when the tilt is bigger than the maximum tilt from the origin.

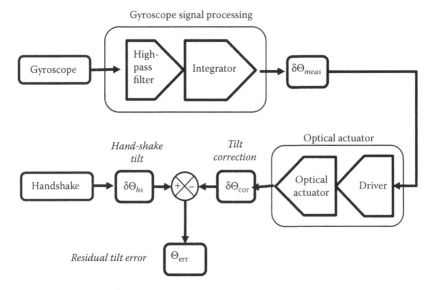

FIGURE 7.13   OIS schematic.

where $F_{gyro}(v)$ and $S_{\Theta gyro}(v)$ are, respectively, the frequency transfer function and the equivalent tilt noise spectral density of the subsystem measuring the hand-shake tilt of the camera. We can simulate the residual hand-shake blur with OIS with the following expression:

$$\Theta x^2_{residual\_OIS}\left(T_{pct}\right) = 4 \cdot \int\limits_{v=0}^{v=\infty} S_{\delta\Theta x_{gyro\_noise}}\left(v\right) \cdot \left|A_{LL}\left(v\right)\right|^2 \cdot \left|A_{T_{pct}}\left(v\right)\right|^2 \cdot dv$$

$$+ 4 \cdot \int\limits_{v=0}^{v=\infty} S_{\delta\Theta x_{hs}}\left(v\right) \cdot \left|1 - F_{gyro}\left(v\right) \cdot A_{LL}\left(v\right)\right|^2$$

$$\cdot \left|A_{T_{pct}}\left(v\right)\right|^2 \cdot dv \qquad (7.16)$$

where $A_{LL}(v)$ is the frequency response of the optical tilt actuator. The first term of the expression gives the noise contribution of the tilt measurement subsystem and the second term estimates the intrinsic residual OIS blur according to the type of hand-shake noise and the frequency response of the tilt actuator. The blur reduction factor (BRF) of the OIS system can be expressed as follows:

$$BRF = \frac{\Theta x_{residual-OIS}}{\delta\Theta x_{rms}} \tag{7.17}$$

The BRF performed by an OIS is sometimes compared to a reduction of the picture integration time or to an aperture stop number gain that would be needed to perform the same blur reduction.[*] The relative reduction of the picture integration time, $gain_T$, is obtained from the hand-shake model in Equation 7.4:

$$gain_T = BRF^{1/0.62} \tag{7.18}$$

and the equivalent number of stop gains is:

$$N = \frac{\ln(BRF)}{0.62 \cdot \ln(2)} \tag{7.19}$$

Example 7.1

We have simulated the performances of an OIS system using a liquid lens with the frequency response described in Equation 7.12, with $m = 0.73$ and $v_0 = 14$ Hz, and with the hand-shake power spectral density model of Equation 7.6 (Simon et al. 2010). We assume the use of a perfect gyroscope with negligible noise and a response $F_{gyro}(v) = 1$.

The result of the simulation presented in Figure 7.14 shows the significant dependence of the blur reduction efficiency of an OIS with the integration time. In this particular example, the blur is reduced by a factor greater than 2 for a picture integration time greater than 100 ms. The BRF reaches a value of 3.5 for a picture integration time of 300 ms corresponding to an equivalent aperture increase of three stops. In other words, performing the same blur reduction efficiency with a larger optical aperture lens would require going from a typical f/2.8 aperture value to f/1. This unrealistic value, with today's miniature optical designs, shows the decisive contribution of OIS to miniature camera performances.

---

[*] In terms of light exposure, a number of aperture stop gains of N is equivalent to an increase of the exposure time by a factor of 2^N. It corresponds to a reduction of the F-number by a factor of $2^{N/2}$.

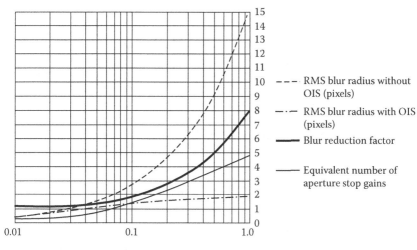

Picture integration time (s)

FIGURE 7.14 Performance evaluation of an OIS system based on the frequency response of the actuator.

References

Ahn, J.-W., Kim, T.-H., Jeong, J.-H., and Suh, J.-G. (2010), Development of subminiature hand shaking compensation device for mobile phone camera system, *Proceedings of the Department of the Korean Society for Noise and Vibration Engineering*, 588–589.

Chiu, C.-W., Chao, P.-C.P., Kao, N.-Y.Y., and Young, F.-K. (2008), Optimal design and experimental verification of a magnetically actuated optical image stabilization system for cameras in mobile phones, *J. Appl. Phys. 103*, 07F136.

Golik, B. and Wueller, D. (2007), Measurement method for image stabilizing systems, *Proc. SPIE, SPIE 6502*, 65020O-1–65020O-10.

Kauhanen, P. and Rouvinen, J. (2006), Actuator for miniature optical image stabilizer, *IEEE Trans Magn, 10th International Conference on New Actuators*, Bremen, 549–552.

Kim, C., Song, M.-G., Park, K.-S., Park, Y.-P., and Song, D.-Y. (2011), Design of a hybrid optical image stabilization actuator to compensate for hand trembling, *Microsyst. Technol. 17*, 971–981.

Kinugasa, T., Yamamoto, N., Komatsu, H., Takase, S., and Imaide, T. (1990), Electronic image stabilizer for video camera use, *IEEE Trans. Consum. Electron. 36* (3), 520–525.

Sachs, D., Nasiri S., and Goehl, D. (2006), Image Stabilization Technology Overview, *Whitepaper*, InvenSense Inc, *Santa Clara*. http://www.invensense.com/mems/gyro/documents/whitepapers/ImageStabilizationWhitepaper_051606.pdf.

Shimohata, T., Tsuchida, Y., and Kusaka, H. (2002), Control technology for optical image stabilization, *SMPTE J. 111*, 609–615.

Simon, E., Berge, B., Fillit, F., Gaton, H., Guillet, M., Jacques-Sermet, O., Laune, F., Legrand, J., Maillard, M., and Tallaron, N. (2010a), Optical design rules of a camera module with a liquid lens and principle of command for AF and OIS functions, *Proc. SPIE. 7849*, 784903.

Simon, E., Berge, B., Gaton, H., Jacques-Sermet, O., Laune, F., Legrand, J., Maillard, M., Moine, D., and Verplanck, N. (2010b), Optical image stabilization with a liquid lens, *Proc. Int'l Conf. Opt.-photon. Design Fabr.*

Simon, E., Craen, P., Gaton, H., Jacques-Sermet, O., Laune, F., Legrand, J., Maillard, M., Tallaron, N., Verplanck, N., and Berge, B. (2010c), Liquid lens enabling real-time focus and tilt compensation for optical image stabilization in camera modules, *Proc. SPIE 7716*, 77160I.

Song, M.-G., Baek, H.-W., Park, N.-C., Park, K.-S., Yoon, T., Park, Y.-P., and Lim, S.-C. (2010), Development of small sized actuator with compliant mechanism for optical image stabilization, *IEEE Trans. Magn. 46* (6), 2369–2372.

Xiao, F., Farrell, J.E., and Wandell, B. (2005), Psychophysical thresholds and digital camera sensitivity: The thousand photon limit, *Proc. SPIE, San Jose, 5678*, 75–84.

Xiao, F., Farrell, J.E., Catrysse, P.B., and Wandell, B. (2009), Mobile imaging: The big challenge of the small pixel, *Proc. SPIE 7250*, 72500K.

Xiao, F., Pincenti, J., John, G., and Johnson, K. (2007), Camera-motion and mobile imaging, *Proc. SPIE 650204*.

Xiao, F., Silverstein, A., and Farrell, J. (2006), Camera-motion and effective spatial resolution, *Proc. SPIE 7241*, 33–36.

Yotam, E., Ephi, P., and Ami, Y. (2007), MTF for Bayer pattern color detector, *Proc. SPIE 6567*, 65671M.

Yu, H.C. and Liu, T.S. (2007), Design of a slim optical image stabilization actuator for mobile phone cameras, *Phys. Stat. Sol. C 4* (12), 4647–4650.

CHAPTER **8**

Multiaperture Cameras

Andreas Brückner

Contents

8.1 Introduction

Vision is by far the most important sense of human beings. It enables exact orientation in a three-dimensional (3D) environment, it reaches up to several kilometers in distance, and it has always been the transport medium of humans' memories. The majority of man-made optical imaging systems follows the design principles of single-aperture optics, which is the basic architecture of all known mammalian eyes [1]. Single aperture refers to the fact that all light that is recorded in an image passes through a single clear aperture of the optical system. Even the ancestor of the photographic camera, the *camera obscura*, which had no lens at all, belongs to this type [2,3]. In the second half of the nineteenth century, the manufacturing of optical glasses of constant quality enabled the spread of mature optical systems such as telescopes, microscopes, and photographic objectives into scientific, medical, and consumer applications [4]. About 100 years later, the fabrication of optoelectronic image sensors using photolithographic techniques known from microelectronics engineering, and the development of precise molding of plastic lenses using high-grade optical polymers paved the way for the miniaturization of imaging systems [5]. Shrinking pixel sizes enabled an increase of the resolution in image space while reducing the image sensor size, and the fabrication of highly precise aspherical lenses led to the development of compact and cheap optical imaging systems. Hence, digital cameras are now an integral part of various electronic products, such as laptops, tablet personal computers (PCs), and mobile phones, automotive safety systems, video endoscopes, and point-of-care diagnostics to name only a few. Even though today's commercially available camera modules range down to $2.5 \times 3 \times 2.5$ mm in size, manufacturers strive for even smaller packages. In particular, the module height is the target for further miniaturization in order to allow its integration into the latest ultraslim products. Currently, the pixel pitch size of the smallest image sensor format is between 1.1 and 1.75 μm. Further downscaling has yielded pixel sizes below the diffraction limit, resulting in a reduced spatial resolution as well as increased noise. Although this might be tolerable in consumer digital cameras, the effect on the performance of subsequent image processing and analysis is not acceptable in machine vision, industrial inspection, and medical imaging. Additionally, technological limits apply, such as tight mechanical tolerances, which greatly increase the costs for a hybrid integration of small lenses. Lower *f*-numbers are hard to achieve, so the diffraction limit cannot be shifted, leading to insufficient performance with standard fabrication techniques [6,7].

For further miniaturization of optical imaging systems, it is worth looking at some of the fascinating approaches that have been present in the tiniest creatures in nature for millions of years. The evolutionary solution of choice is found in the vision systems of invertebrates—the compound eye. Here, a large number of tiny vision systems (called ommatidia) on a curved base capture the visual information of a large field of view (FOV) in parallel at the cost of spatial resolution. Each ommatidium has a small diameter and a low information capacity when compared to the single-aperture eye. However, due to the large number of channels, a high information capacity of the overall multiaperture objective can be achieved.

Since the dawn of digital image sensors, several technical derivatives of compound eyes have been realized in order to miniaturize man-made vision systems [8–14]. However, as the major challenge to technically adapting natural compound eyes is the required precision of their fabrication and assembly, and since macroscopic fabrication methods have been exploited for the manufacturing of microoptical structures, none of the attempts has so far led to a successful technology. The first examples of technically derived compound eyes by well-adapted microoptical technologies provided the thinnest known optical sensors, which were promising even though the achieved resolution was not yet appropriate for technical applications [15–20].

The aim of this chapter is to demonstrate the promising approach of the combination of the following:

1. Bio-inspired system design principles: The general principles of insect vision such as the simplification of the design, the spatial segmentation of the FOV, the spectral segmentation,

and the overlap of the FOVs of adjacent optical channels are applied to yield new approaches to miniaturized vision systems with a high resolution, which overcome the basic scaling limits of single-aperture imaging optics.

2. High-precision microoptics molding technology: The discussion is focused on optical systems that can be fabricated in parallel by lithography and ultraviolet (UV)-molding technology based on wafer-level techniques which are suitable for high-volume production at low cost.

3. Real-time digital image processing: This is a key element for successful application as the image information is often loosely scattered across the image plane in a multiaperture imaging system.

The chapter starts with an explanation of the limits on the miniaturization of conventional camera optics, which are currently encountered in the slimmest consumer electronic devices on the market. A few basic properties of single-aperture imaging optics are reviewed in order to support the discussion before the scope focuses on multiaperture imaging optics as a new and competitive way to achieve the slimmest camera devices. After giving further insight into why multiaperture optical systems can lead to slimmer setups, a novel classification is introduced which distinguishes three different kinds of multiaperture imaging systems according to their segmentation of the FOV. In addition to the well-known *apposition type* [9,15], the electronic and optical stitching of segments are of central interest. The electronic stitching creates a digital image by the post-processed fusion of images of FOV segments that have been captured in separate optical channels, whereas the optical stitching of segments achieves a regular image of the object in the image plane prior to capture. A fourth type—the multiaperture *super-resolution*—is added for completeness. It differs from the others as it uses no segmentation of the field. The discussion includes a number of example demonstrations of these kinds of multiaperture optical imaging systems. One of the examples examined in detail is the *electronic cluster eye* (eCLEY), which uses an overlap of the FOVs of adjacent optical channels with a subpixel displacement in order to increase the sampling in the object plane. Its potential for a high resolution makes it the most promising candidate for miniature camera applications. Different examples with respect to size and resolution are investigated whereas two demonstration systems have been realized so far using microoptical fabrication at wafer level. The software stitching is discussed together with the correction of distortion by image processing as well as the prospects of *eCLEYs* with respect to megapixel resolution.

8.2 Fundamentals

After more than 170 years of photographic lens history, the properties, tradeoffs, and mathematical ways of calculating single-aperture objective lenses are well understood [3,21–27].

In the recent past, *wafer-level optics* (WLO) evolved from microoptics technology as an alternative fabrication technique and it has been established for the fabrication of low-end camera modules with lenses of <3 mm in diameter [28,29]. A major advantage over injection molding is that the fabrication of lenses and the integration with spacers to objectives is carried out at wafer level for thousands of lenses in parallel. Additionally, processes are operated in a clean-room environment so that contamination can be limited. However, the adoption of UV-molding techniques from the world of microoptics poses several challenges for the replication of millimeter-scale lenses with large sags. Mold wafer mastering accuracy, material shrinkage, and tight z-height tolerances for the whole objective are issues that place a strong limitation on the performance and yield of WLO cameras. To date, commercially available WLO solutions hardly achieve a resolution higher than a video graphics array (VGA) (640 × 480 pixels). Figure 8.1a shows a state-of-the-art miniature objective lens design, which is especially optimized for production with WLO technology. This simple example will be used in order to derive some basic parameters that are needed for the discussion.

The effective focal length, f_{eff}, is the central parameter of an objective lens. With the aberration-free, thin lens approximation, it may be derived from Figure 8.1b using the diagonal of the image sensor (D_{im}) and the half angle of the diagonal FOV (α, $n = 1$ in object and image spaces):

$$f_{\text{eff}} = \frac{D_{\text{im}}}{2 \cdot \tan(\alpha)}. \tag{8.1}$$

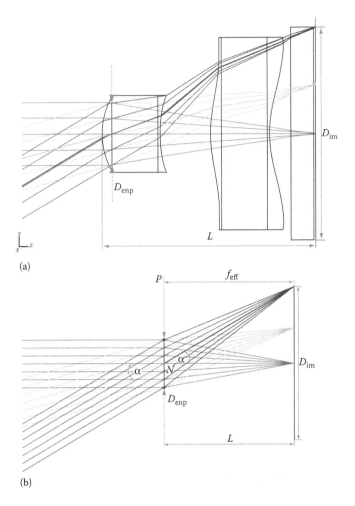

FIGURE 8.1 (a) Layout plot of a state-of-the-art miniature lens design example, which contains two convex–concave components with aspherical surfaces. The cover glass of the image sensor is placed in front of the image plane. (b) Thin lens or a paraxial approximation of the lens from (a). Additional annotations show: the half angle of the full field of view α, the diameter of the entrance pupil D_{enp}, the total track length L of the lens, the effective focal length f_{eff}, and the diameter of the image circle D_{im}. The major principal plane and the nodal point are denoted by P and N, respectively.

The *f*-number in image space is calculated by [27]

$$F/\# = \frac{f_{\text{eff}}}{D_{\text{enp}}} \tag{8.2}$$

in the paraxial approximation and for an object at infinity.

8.2.1 Resolution

A central figure of merit of an imaging system is the resolution or the ability to form distinguishable images of significantly close points in the lateral dimension of object space. This discussion will distinguish between *diffraction-limited* and *aberration-limited* resolutions. The term *diffraction limited* means that, viewed from a ray-optical standpoint, the point image of a single in-focus object point is smaller than the diffraction blur, which states the minimum physical spot size caused by diffraction at the finite aperture stop of the imaging system after a wave-optical model. In the *aberration-limited* case, the image spot size of an in-focus object point is larger than the diffraction blur due to insufficiencies of the imaging optics. A general description of the resolution is provided by the *point spread function* (PSF) of an imaging system [30].

If aberrations are negligible, the size of an image point equals the *diffraction-limited PSF* that, for a lens with a circular aperture, is often referred to as the normalized *Airy diffraction pattern* [3,30,31]. The diameter of the center lobe is the *Airy diameter*, which states the physical limit for the size of a single image point [3,25,27,30]:

$$d_{\text{Airy}} = 2.44 \cdot \lambda F/\#. \tag{8.3}$$

The effective size of an image point, A_p (sometimes referred to as two-dimensional [2D] blur) may be approximated by the sum of the Gaussian moments of diffraction and transverse aberrations, δ_a [32], which gives

$$A_p = \left(\delta_{\text{diff}}\right)^2 + \left(\delta_a\right)^2 = \left(\lambda F/\#\right)^2 + \left(\delta_a\right)^2. \tag{8.4}$$

Moreover, the smallest resolvable angle in object space, $\Delta\delta$, results from the projection of Equation 8.4 with the focal length:

$$\Delta\delta = \sqrt{\left(\frac{\lambda}{D}\right)^2 + \left(\frac{\delta_a}{f}\right)^2}. \tag{8.5}$$

The *space-bandwidth product* (SBP), which is defined as the number of resolvable image points over the image area A_I, is a useful parameter for comparing imaging systems [32]:

$$\text{SBP} = \frac{A_{\mathrm{I}}}{\left(\lambda F/\# \right)^2 + \left(\delta_{\mathrm{a}}\right)^2}. \tag{8.6}$$

The SBP is also denoted as the spatial information capacity of an imaging system.

In the practical case, imaging optics are rarely diffraction limited and the correction of aberrations plays a major role in the design and optimization process in order to achieve an adequate resolution and image quality. As the specific amount of aberrations reveals little about the resolution of an imaging system, a common way of assessing the latter is introduced by analyzing the spatial frequency response of the system.

8.2.2 Modulation Transfer Function

The *modulation transfer function* (MTF) is widely used for the characterization of the resolution of imaging optics because a complex optical setup can be summarized in a few MTF curve diagrams. The MTF is directly linked with the PSF via a Fourier transform [30].

The MTF describes the contrast in the image plane relative to the contrast in the object plane, depending on the spatial frequency of the input modulation. Thus, it can be measured by determining the contrast, K, of a certain spatial frequency, f_x, from the minimum and maximum intensities, $E_{\min}$ and $E_{\max}$, in the image:

$$K\left(f_x\right) = \frac{E_{\max}\left(f_x\right) - E_{\min}\left(f_x\right)}{E_{\max}\left(f_x\right) + E_{\min}\left(f_x\right)} \tag{8.7}$$

and normalizing it to the contrast of the equivalent frequency component in the object plane [27]:

$$\hat{H}\left(f_x\right) = \frac{K_{\mathrm{im}}\left(f_x\right)}{K_{\mathrm{ob}}\left(mf_x\right)}. \tag{8.8}$$

For a given optical imaging system, the MTF generally varies with the position in the image plane (variation with field angle), with wavelength, and also with orientation in the spatial frequency plane.

8.2.3 Optical Cutoff Frequency

The optical cutoff frequency, ρ_0, indicates the spatial frequency at which the MTF vanishes and thus zero contrast is obtained in the image independent of the object contrast.

The cutoff frequency, ρ_0^{diff}, of the diffraction-limited system is linked to the physical properties of the lens such as the radius of the (circular) clear aperture, $r_a = D_{\text{enp}}/2$, the wavelength of light, λ, and the axial position of image formation, z_{im}, which is equal to the focal length, f, in the case of infinite object distance [27,30]:

$$\rho_0^{\text{diff}} = \frac{2r_a}{\lambda z_{\text{im}}} \overset{z_{\text{im}}=f}{=} \frac{1}{\lambda F/\#}. \tag{8.9}$$

The influence of certain aberrations on the MTF performance can be studied in Smith [27], Goodman [30], and Mahajan [31].

8.2.4 Depth of Focus

The depth of focus is defined as the axial distance of defocus (δ_I), which causes a blur diameter (b_I) under a certain tolerance threshold in the image space. It is closely related to the depth of field (δ_0), which is the axial distance in object space that is sharply imaged within this tolerance [27].

For short focal lengths or large f-numbers, the longitudinal diffraction blur is a good approximation for the depth of focus [27,32,33]:

$$\delta_I^{\text{diff}} = 4\lambda \left(F/\# \right)^2 \tag{8.10}$$

and using relations for the depth of field and the magnification, it follows for the depth of field caused by diffraction blur that [34]

$$\delta_0^{\text{diff}} = \frac{4\lambda \left(F/\# \right)^2 \cdot (s+f)^2}{f^2} = 4\lambda \left(F/\# \right)^2 \cdot \left[\left(\frac{s}{f} \right)^2 + 2\frac{s}{f} + 1 \right]. \tag{8.11}$$

8.3 Scaling Limits of Single-Aperture Imaging Optics

This section discusses the differences that result for conventional imaging optics of different scales. The effects of scaling on the properties of the imaging system that were discussed in the previous sections will be studied. Therefore, two different ways of scaling are distinguished: (1) scaling with a constant f-number and (2) scaling with a constant SBP.

8.3.1 Scaling with Constant f-Number

The focal length, the lens diameter, and the image circle diameter are multiplied by the same factor, M, which yields a linear scaling of the whole camera (Figure 8.2b) [35]. The ratio between the focal length and the aperture diameter is constant (Equation 8.2), which gives a constant FOV (2α) according to Equation 8.1.

With a reduced size, the aberrations decrease linearly with M [32], leading to the *diffraction-limited spot size* of Equation 8.3 and a diffraction-limited MTF as shown in Figure 8.3.

However, Equations 8.9 and 8.3 indicate that scaling with a constant f-number does not influence either the system's cutoff frequency or the smallest possible spot size. Hence, the best possible image resolution is independent of scale [32]. In the diffraction-limited case, the number of resolvable image points (or the SBP) decreases according to M^2 as the image diagonal is decreased. This can be interpreted as a reduction of the angular resolution and thus a scaling of the smallest resolvable angle according to $1/M$ because a larger patch of object space is imaged on the same image point (set $D \to MD$ and $\delta_a/f \to M\delta_a/Mf \to$ constant in Equation 8.5).

According to Equation 8.10, the depth of focus stays constant but it translates into a different depth of field, since δ_0^{diff} depends on $(s/f)^2$. The nominal object distance, s, is fixed by the application, which means that the depth of field grows indirectly proportional to $(Mf)^2$ (see Equation 8.11).

In the case that the pixel size, p_{px}, is scaled together with the image circle diameter, the Nyquist frequency, v_{Ny}, is shifted with respect to the diffraction-limited MTF, which means that for smaller pixel sizes also, the contrast at the Nyquist frequency drops, as shown in Figure 8.3:

$$p_{px} \to Mp_{px} \quad v_{Ny} \to v_{Ny}/M. \tag{8.12}$$

For larger values of M, the system is limited by undersampling and aliasing will occur. For smaller M, the optics MTF will reach a noise level below the Nyquist frequency and the system

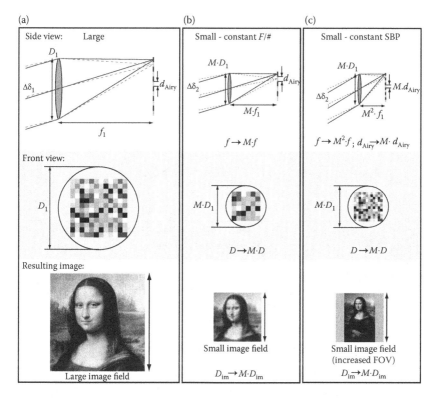

FIGURE 8.2 Scaling of a single-aperture lens assuming diffraction-limited imaging. (a) Large lens: The size of one image point is proportional to $\lambda F/\#$. Due to the large focal length f_1, the angular projection of an image point into object space $\Delta\delta_1$ is small. The lens has a high angular resolution and a large space-bandwidth product (SBP) caused by the extension of the image circle and the small spot size d_{Airy}. (b) Small lens with the same $F/\#$: The size of one image point is the same as in (a). However, due to the shorter focal length, the angular projection of one spot $\Delta\delta_2$ is enlarged. Hence, the lens has a lower angular resolution and a reduced SBP due to the smaller image circle. (c) Small lens with the same space-bandwidth product (SBP): The spot size is decreased due to the reduction of the $F/\#$. The same angular resolution as in (b) results as the focal length is decreased with the size of one image point. Consequently, the information capacity in the image plane equals that of the large system but the information is gathered from a larger field of view with a reduced angular resolution. (Adapted from Völkel, R., Eisner, M., and Weible, K. J., *Microelectronic Engineering*, 67–68, 461–472, 2003.)

will be limited by optical blur. Thus, the information capacity of the image sensor pixel could not be exploited due to oversampling. Hence, downscaling by M does not strictly mean a scaling of the angular resolution by $1/M$, but the angular resolution will decrease additionally according to the loss of contrast for the higher spatial frequencies that necessarily have to be transferred by the miniaturized system.

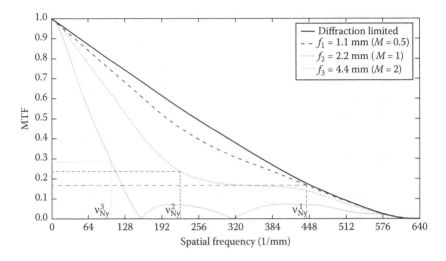

FIGURE 8.3 Monochromatic MTF ($\lambda = 550$ nm) in the image center simulated for three different scales of the objective lens of Figure 8.1a with dashed curve $M_1 = 0.5$, dotted curve $M_2 = 1$, and gray curve $M_3 = 2$. The solid black curve shows the diffraction-limited MTF for $F/\# = 3.0$.

8.3.2 Influence of Scaling on Light Collection Ability

The light collection ability of the optical system can be maintained during scaling with a constant f-number. Even with an extremely decreased aperture stop diameter, the loss of the angular resolution counterbalances that of the light collection ability. Light that is imaged on a given image point is collected from a larger cone angle in the object space [36].

The radiant power that is incident on one pixel in the image plane can be calculated by [15,23,25,33]

$$P_{im} = \pi\tau \cdot B_{ob} \cdot \frac{d^2}{4 \cdot \left(F/\#\right)^2}, \tag{8.13}$$

where B_{ob} is the radiance (the radiant power per surface area element and steradian) emitted by the distant object surface, the transmission coefficient of the optical system τ, and the photosensitive area $A_{ps} = d^2$ of a squared pixel. If the pixel size is scaled by M, the number of photons and thus the radiant power in one pixel is scaled by M^2. The signal-to-noise ratio (SNR) of the image sensor decreases because noise and pixel cross-talk increase with the shrinking pixel size [37]. For example, it can be shown that for indoor video the minimum pixel size for a still acceptable SNR is already reached in current image sensors of the smallest pixel pitch [38]. A further decrease in pixel size is unlikely, although noise is increasingly treated by signal post-processing.

8.3.3 Scaling with Constant SBP

In theory, there is another option, which is the scaling of imaging systems with a constant SBP, illustrated in Figure 8.2c. Here, the smallest resolvable image point, A_P, has to be scaled by M^2 in order to compensate for scaling down the image area, A_I, in Equation 8.6. As the aberrations, δ_a, scale linearly with M, the *f*-number has to be changed by M, which results in a scaling of the focal length, *f*, by M^2. However, the scaling will alter the FOV ($\alpha \rightarrow \alpha/M$) due to the reduced focal length in Equation 8.1. The miniaturized system could achieve the same SBP as a large optical system, but it transfers a larger number of the same coarse object patches as in Figure 8.2b. The angular resolution is also decreased in comparison with case (a) as a result of the scaling process. Furthermore, the maximization of the SBP by scaling up the FOV and decreasing the *f*-number is futile due to the dependency on the aberrations (SBP $\propto 1/\delta_a^2$), which will increase with the growing FOV and the shrinking *f*-number [27]. Scaling down the *f*-number has several additional effects such as a reduced depth of focus, a shift of the diffraction-limited MTF, and an increased light collection ability of the optical system. However, in reality, the fabrication of a small objective lens with a low *f*-number and a large FOV is a very challenging task as it asks for a series of very high-precision surfaces with small radii of curvature, large sags, and dramatically decreased tolerances. Thus, the application of scaling with a constant SBP is very limited for current technological solutions.

8.3.4 Summary of Scaling of Single-Aperture Imaging Optics

In conclusion, the miniaturization of a single-aperture optical system has to be offset by a reduction in both the information capacity (or SBP) and the angular resolution when scaling with a constant *f*-number and FOV. For a large optical imaging system ($M \gg 1$), the SBP and the smallest resolvable angle approach constant values whereas the contribution of diffraction vanishes. The system is aberration limited and any changes of the absolute values for the SBP and the angular resolution have to be achieved by increasing the system's complexity (e.g., adding additional lens elements) in order to decrease the amount of aberrations. For the small scales ($M < 1$), the behavior of a diffraction-limited system is approached, with a decreasing SBP according to M^2 and a decreasing angular resolution according to $1/M$. A scaling down with a constant SBP requires a system with a decreased f-number and a larger FOV, which is not feasible with respect to the technological limits. A certain minimum system size is needed for both a high number of image details and a high angular resolution of a single-aperture optical system [32].

8.4 Basic Parameters of Multiaperture Imaging Optics

The evolutionary solution of choice for the scaling problems of the preceding section is found in the vision systems of invertebrates—the compound eye. Here, a large number of tiny vision systems (called ommatidia) on a curved basis capture the visual information of a large FOV in parallel. The use of multiple optical channels breaks the tradeoff between the focal length and the size of the FOV. Although each ommatidium exhibits a small FOV and thus a small information capacity, the sum of all ommatidia of the compound eye provides a large FOV and an information capacity that is large enough to enable accurate and fast navigation through the insect's habitat.

In order to formulate a general description of planar multiaperture optical imaging systems (MAO), it is noted that every parameter that is applicable for a whole single-aperture optical imaging system (SAO) is also applicable for each individual channel of the MAO. However, beyond that, there are additional parameters that are special to multiaperture setups and these will be explained in the following section.

8.4.1 Basic Geometric Parameters

A multiaperture optical system is structured into a two-dimensional array of $K_x \times K_y$ individual optical channels. An optical channel is defined as the axial assemblage of at least one aperture stop together with at least one focusing element (e.g., a lens) and a spacer for the related propagation length to the image plane. Each channel is associated with a two-dimensional partial image of $n_x \times n_y$ pixels. In the case of a square partial image, the number of pixels is $n_x = n_y = n_g$. The physical edge length of the square partial image is

$$I_p = n_g \cdot p_{px}. \tag{8.14}$$

The number of pixels in the final image is defined by N_x and N_y along the x- and y-direction. This corresponds directly to the number of object sampling intervals (see Figure 8.4a). Thus, the number of optical channels (in x and y) is related to the ratio of the total pixel number and the number of pixels per partial image using

$$K_{x,y} = \frac{N_{x,y}}{n_{x,y}}. \tag{8.15}$$

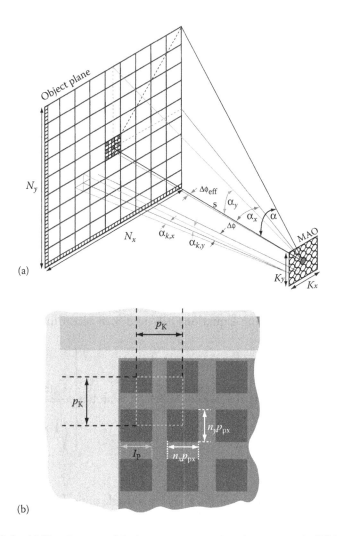

(a)

(b)

FIGURE 8.4 (a) Visualization of the basic parameters that characterize the FOV of multiaperture optics (MAO). (b) Schematic overview of the basic parameters in the image plane of multiaperture optical systems.

For simplicity, it is assumed that the $K_x \times K_y$ partial images are arranged in a regular Cartesian array with equal pitch, p_K, along the x- and y-direction (see Figure 8.4b). Hence, a one-dimensional (1D) optical fill factor, Γ, in the image plane is defined as

$$\Gamma = \frac{l_p}{p_K}. \tag{8.16}$$

The 2D optical fill factor is given by Γ^2 in the case of quadratic partial images in a square array lattice.

8.4.2 FOV

In most cases, the diagonal FOV (2α) is determined by its application. From rectangular triangles and trigonometric relations, one relates α to the half-cone angles of the FOV along the horizontal and vertical directions α_x and α_y:

$$\tan^2 \alpha = \tan^2 \alpha_x + \tan^2 \alpha_y, \tag{8.17}$$

and the ratio of the horizontal to the vertical edge length of the image is found using

$$\frac{\tan \alpha_x}{\tan \alpha_y} = \frac{N_x}{N_y}. \tag{8.18}$$

The half angles of the horizontal and vertical FOVs may be calculated using Equations 8.17 and 8.18 according to

$$\alpha_x = \arctan\left[\frac{\tan(\alpha)}{\sqrt{1+\left(\frac{N_y}{N_x}\right)^2}}\right]; \alpha_y = \arctan\left[\frac{\tan(\alpha)}{\sqrt{1+\left(\frac{N_x}{N_y}\right)^2}}\right]. \tag{8.19}$$

The half angles of the FOV for a single optical channel in x, y are

$$\alpha_{K,x} = \arctan\left(\frac{n_x \cdot p_{px}}{2f}\right); \alpha_{K,y} = \arctan\left(\frac{n_y \cdot p_{px}}{2f}\right), \tag{8.20}$$

which reduce to a symmetric FOV per channel in the case of a square partial image.

8.4.3 Sampling Angle

Following the physiological definitions, the *interommatidial angle* is the angular offset between the central viewing directions (optical axes) of adjacent optical channels [9,36]:

$$\Delta\phi = \arctan\left(\frac{\Delta p_K}{f}\right). \tag{8.21}$$

Two different sampling angles are defined in addition to the interommatidial angle: the interpixel angle, $\Delta\phi_{px}$, which is the offset angle between two adjacent pixels of a single channel when projected in object space:

$$\Delta\phi_{px} = \arctan\left(\frac{p_{px}}{f}\right), \tag{8.22}$$

and the effective sampling angle:

$$\Delta\phi_{eff} = \arctan\left(\frac{\Delta p_{eff}}{f}\right), \tag{8.23}$$

which is the smallest angular sampling interval in object space. It could be created by the effect of a pair of different optical channels with an effective pitch difference of Δp_{eff}, which is discussed later in the chapter.

The total number of sampling points is limited by the number of pixels in the final image, and the effective sampling angles in x and y can be alternatively calculated for an equidistant sampling of the FOV:

$$\Delta\phi_{eff,x} = \arctan\left(\frac{\tan\alpha_x}{N_x/2}\right); \Delta\phi_{eff,y} = \arctan\left(\frac{\tan\alpha_y}{N_x/2}\right). \tag{8.24}$$

8.4.4 Interchannel Distortion

When imaging a planar object onto a planar image surface, an equidistant sampling in angular space leads to a nonequidistant

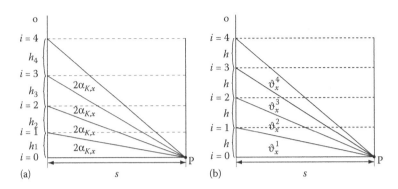

FIGURE 8.5 Sampling schemes with an equidistant angle in FOV (a) and an equidistant distance on the object plane (b). The object plane is denoted by O. The optical system in P is placed at a distance s.

sampling on the object plane (see Figure 8.5a) and thus a distorted image. However, the viewing direction of the individual channels in a multiaperture system can be chosen arbitrarily. Hence, they can be arranged in such a way that a regular grid on a planar object surface is formed (see Figure 8.5b). This is achieved if the viewing angle, ϑ_r, of the channel with indices (i,j) along the horizontal and vertical dimensions in the array follows the equation [39]:

$$\vartheta_r = \arctan\left(\tan(\alpha) \cdot \frac{\sqrt{(i)^2 + (j)^2}}{\sqrt{\left(\frac{K_x - 1}{2}\right)^2 + \left(\frac{K_y - 1}{2}\right)^2}} \right) \quad (8.25)$$

or for a separate calculation along the x- and y-direction:

$$\begin{pmatrix} \tan\vartheta_x \\ \tan\vartheta_y \end{pmatrix} = \frac{\tan(\alpha)}{\sqrt{\left(\frac{K_x - 1}{2}\right)^2 + \left(\frac{K_y - 1}{2}\right)^2}} \cdot \begin{pmatrix} i \\ j \end{pmatrix}. \quad (8.26)$$

The numbers of channels K_x and K_y are assumed to be odd without loss of generality and $i,j \in [0;(K_{x,y} - 1)/2]$.

The easiest way to achieve such a correction of *interchannel distortion* is to adapt the local pitch difference according to the position of the channel in the array in Equation 8.21. The result is an irregular or so-called chirped microlens array (MLA) [39–41].

8.5 Reasons for Thickness Reduction in MAO

There are a few fundamental properties that summarize the advantages of multiaperture imaging optics in terms of the miniaturization of imaging systems:

> *Spatial Segmentation of the FOV*: Each part of the total FOV that is transferred by an individual optical channel originates from a different position in object space. Thus, the properties of each optical channel can be adapted in order to create the best possible image of a particular segment. A smaller range of field values has to be imaged by each individual channel so that the correction of aberrations becomes much easier because most aberrations strongly depend on field coordinates. The resulting advantages with respect to single-aperture optics are a simple optical setup with few elements within each channel, a relatively small diameter and height for each of the microlenses, and the feasibility of a large total FOV.
>
> *Spectral Segmentation*: The information that is transferred by the individual optical channels differs in its spectral composition. Each channel transfers a part of the full spectral range of the multiaperture imaging system. This feature eases the correction of chromatic aberrations for the limited spectral range of the individual channel, which yields a more simple optical setup with fewer elements within each channel in comparison with single-aperture imaging optics.
>
> *Overlap of the FOVs of Different Optical Channels*: The FOVs of different optical channels overlap to a desired extent, which allows for features that are impossible for single-aperture imaging optics. Some examples are the decrease of the total track length of the optical module due to increased sampling, the increase of light collection ability and information capacity (color, multispectral, and polarization imaging), as well as 3D imaging.

8.6 Concepts for FOV Segmentation

The segmentation of the FOV by multiple optical channels can be carried out in several ways, which lead to the different basic architectures of the multiaperture optical systems (MAOs) that

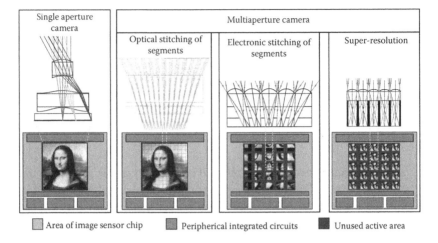

FIGURE 8.6 Overview of the principles of single and different multiaperture optical imaging systems. (From https://en.wikipedia.org/wiki/File:Mona_Lisa,_by_Leonardo_da_Vinci,_from_C2RMF_retouched.jpg.)

are shown in Figure 8.6. The following discussion provides a classification that allows the various principles of the state-of-the-art and the novel approaches for multiaperture imaging systems to be distinguished.

8.6.1 Apposition Type (APCO)

The artificial apposition compound eye (APCO) uses only one pixel per optical channel, as illustrated in Figure 8.7b. Thus, $n_g = 1$ and the size of a partial image is that of a single pixel defined by either the pixel pitch, p_{px}, or the diameter of the photodiode d in the case that both differ. Consequently, the number of pixels in the final image is directly given by the number of channels in Equation 8.15 and the optical fill factor is $\Gamma = d/p_K$. The apposition type is mentioned separately here for the sake of completeness, although it can be counted as a special case of the class of electronic stitching of segments. If the size of the photodiode d is comparable to the size of the PSF, the full FOV per channel ($2\alpha_{K,x}$ and $2\alpha_{K,y}$) equals the acceptance angle $\Delta\phi$. It defines the finite angular interval centered on the corresponding optical axis in that the photoreceptor of each channel accepts incoming light [42,43]:

$$\Delta\varphi = \sqrt{\left(\frac{d}{f}\right)^2 + \left(\frac{\lambda}{D}\right)^2}. \tag{8.27}$$

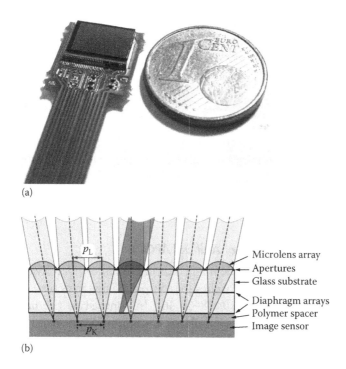

(a)

(b)

FIGURE 8.7 (a) Ultrathin imaging sensor formed by an artificial apposition compound eye (APCO) attached to a customized CMOS image sensor on a flexible printed circuit board. (b) Schematic section of an APCO. The darker-shaded ray bundle displays cross talk that is suppressed by horizontal diaphragm arrays.

The effective sampling angle $\Delta\phi_{eff}$ equals the interommatidial angle $\Delta\varphi$ of Equation 8.21. If the condition

$$\Delta\phi_{eff} = \frac{\Delta\varphi}{2} \qquad (8.28)$$

holds, the maximum transferable angular frequency in a single channel can be reconstructed in the final image [15]. The final image of the APCO can be made free of distortion by correcting the local pitch difference, Δp_K, in order to achieve a sampling according to Equation 8.25 [41].

The APCO defines an extreme case which completely decouples the relationship between the focal length, the FOV, and the image size known from Equation 8.1, because the overall image size of the APCO is determined by the moiré magnification, which is approximately [44,45]:

$$m_{\text{moire}} = -\frac{p_K}{\Delta p_K},$$

(8.29)

and is thus independent of the focal length. It creates an image with a much larger magnification than that of each single optical channel and, additionally, it has a much shorter system length than a single-aperture optical system with the same magnification [15,16].

8.6.1.1 Examples of Artificial APCOs

After optoelectronic image sensors became widely available, various technical approaches for compact vision systems exploited the principle of APCOs [8–10,46]. In contrast to the natural solution, most APCOs are planar configurations because the current image sensor technology is limited to planar substrates. The published solutions mainly suffered from inaccurate assembly because macroscopic fabrication technology was used and the individual arrays were composed bit by bit. The resulting resolutions were too low for any camera application. Later developments led to 3D microfabrication methods, which yielded artificial compound eye optics on a spherical basis [18,20,47,48]. However, the problem of recording the image from the curved image field has remained unsolved in these approaches so far. The application of piezoelectric and thermal actuators was examined for the time-sequential increase of the sampling in object space [11,49] or the expansion of the observed field size including a focus compensation [50]. Although moving parts added a benefit to the system performance, the approaches lacked considerable miniaturization due to the problem of integration with the optical components.

An example of a realized APCO, which was manufactured with microtechnology, consists of a planar MLA on a transparent substrate integrated on an optoelectronic sensor array [15,16]. The thickness of the substrate is matched to the focal length of the individual microlenses, f, so that the detector pixels are located in the focal plane. The difference

$$\Delta p_K = p_L - p_K$$

(8.30)

between the pitch of the lenses, p_L, and the pixels, p_K, is used to achieve different viewing directions for the individual channels. Thus, each channel corresponds to one field angle in object space with the optical axis directed outward (Figure 8.7b), just like in the curved natural APCO [51].

The angular position of contrast edges or point stimuli can be measured with an accuracy that is an order of magnitude higher than the coarse angular resolution, which results for a short focal length [52,53]. Similar to hyperacuity in insect compound eyes [54,55], this is enabled by a controlled overlap between the acceptance angles of adjacent optical channels of an APCO.

Unlike the curved natural APCO, the planar MLA suffers from aberrations under an oblique angle of incidence, which decreases the image resolution with the increasing angle in the FOV. However, as each channel images only a small angular portion of the FOV, its lens parameters can be tuned according to the center viewing angle of the specific channel. Such a *chirped MLA* is formed by, for example, altering the radii of the curvature in the main axes and the local pitch in order to correct for astigmatism, field curvature, and distortion [39–41].

Another advantage of the APCO is the large depth of field. As a result of the short focal length of the microlenses and the use of a single pixel per channel, the image remains sharp independently of the object distance [36].

A series of prototypes of microoptical APCOs was demonstrated based on the aforementioned principles [15,16]. The latest system, which has been developed for out-of-position detection in automotive applications, is shown in Figure 8.7a [51]. Its specifications are given in Table 8.1.

TABLE 8.1 Selected Parameters of the Ultrathin Imaging Sensor from Figure 8.7a

Parameter	Value
Number of channels	144×96
Total track length (L) (μm)	300
Field of view	$85° \times 51°$
F/#	4
Microlens diameter (μm)	50
Pixel size (μm)	3
Size of sensor head (mm³)	$10 \times 10 \times 1.2$
2D fill factor (%)	0.34

The pixel pitch, p_{px}, of the detector array has to be adapted to that of the microlenses, p_L (typical 50–150 μm), while the photodiodes of each pixel should be small (2–10 μm), which is unusual for commercially available image sensors. For this reason, only a fraction of the whole number of pixels is illuminated and read out for large-sized image sensors. Alternatively, a customized image sensor array can be used which additionally offers the possibility of implementing signal preprocessing within the periphery of each pixel. In both cases, a large chip area is consumed by the APCO in order to achieve a certain image resolution. The low image resolution and the increased costs for an image sensor of adequate footprint size are the main problems of the APCO as far as commercial applications are concerned.

8.6.2 Electronic Stitching of Segments

The second class of multiaperture optics uses multiple pixels per channel, which are acquired in parallel within the separated partial images of each individual channel (see Figure 8.6). Subsequently, the distributed image information is fused to a final image of the whole FOV via image processing (software or electronic image stitching).

This setup enables the highest degree of freedom and it will be distinguished between three different sampling schemes:

Redundant Sampling: Light that originates from each object point is sampled in multiple optical channels. This is achieved by choosing a pitch difference between adjacent optical channels, Δp_K, that is the m_0-th fraction of the pixel pitch p_{px} with m_0 being an integer value. Thus, starting from a channel with index l, there is another pixel which captures the same point of a distant object in the channels with indices $l + m_0$ and $l - m_0$ along one row of the MLA (see Figure 8.8). The smallest sampling interval in image space is then given by $\Delta p_{eff} = p_{px}/m_0$. Redundant sampling can be used in order to acquire additional information about the object, such as color, polarization, and multispectral imaging, or to increase the light sensitivity of artificial compound eyes. This sampling is used by the *artificial neural superposition eye* (ANSE), which reveals the inspiration from the neural superposition eye of insects [56].

Increased Sampling: The sampling is improved by overlapping the FOVs of adjacent optical channels. The adjacent

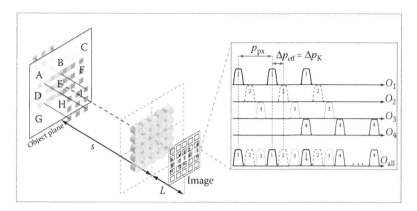

FIGURE 8.8 Illustration of the redundant sampling scheme. On the right, each line O_i shows the sampling along a section in image space of the ith channels for $m_0 = 3$. The last line O_{all} illustrates the final configuration in the post-processed image. The sampling interval of each pixel is indicated by a boxlike function with the index of the channel.

sampling grids need to be displaced by a subpixel amount, Δp_{eff}, that is, the kth fraction of the pixel pitch p_{px}, where k must not be an integer divisor of the number of pixels per channel n_g (Figure 8.9). The main purpose of this type is to decrease the total track length while maintaining the same image resolution compared to single-aperture imaging systems with the same pixel size. The related device that achieves such an *increased sampling* is called the electronic cluster eye (eCLEY). The eCLEY is inspired by the compound eyes of *Strepsiptera* [57,58].

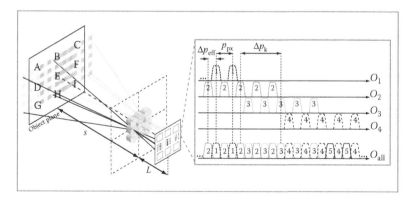

FIGURE 8.9 Illustration of the increased sampling scheme. On the right, each line O_i shows the sampling along a section in image space of the ith channels for $k = 2$ and $n_g = 5$. The last line O_{all} illustrates the final configuration in the post-processed image. A pixel fill factor of about 50% is assumed.

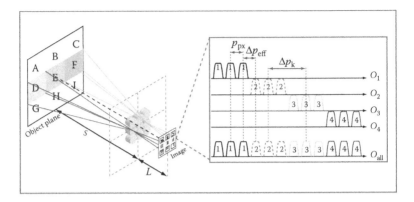

FIGURE 8.10 Illustration of the simple stitching scheme. On the right, each line O_i shows the sampling along a section in image space of the ith channels for $n_g = 3$. The last line O_{all} illustrates the final configuration in the post-processed image.

Simple Stitching: The channel pitch difference, Δp_K, equals the size of the partial images, I_P, and the individual FOVs of adjacent optical channels are seamlessly stitched by post-processing (see Figure 8.10). The effective pitch difference, Δp_{eff}, is equal to the pixel pitch, p_{px}, in each channel. The application of this type can be compared to that of optical image stitching (discussed in the next section) with the difference that it offers the possibility to post-process the partial images prior to image stitching. There is no known natural archetype for the *simple stitching* scheme.

The fraction of image sensor area that is used for the acquisition of partial images for the electronic stitching of segments is given by the 2D optical fill factor:

$$\Gamma^2 = \frac{I_P^2}{p_K^2}. \tag{8.31}$$

There is a gap of $p_K - n_g \cdot p_{px}$ between adjacent partial images, which cannot be used for active pixels so that $\Gamma^2 < 1$ (see Figure 8.4b). Thus, the principle of the electronic stitching of segments asks for a custom image sensor layout.

8.6.2.1 Thickness Reduction Due to Increased Sampling

When shrinking the size of the optical system, the interpixel angle $(\Delta \phi_{px})$ in a single-aperture system increases with the decreasing focal length according to Equation 8.22. However, the viewing direction in the FOV can be freely chosen by applying the pitch difference (Δp_k) in the case of multiaperture imaging optics with the electronic stitching of segments. Hence, the angular sampling of the object space is not limited by the pixel pitch and the focal length of a single channel, but is limited by the effective pitch difference Δp_{eff} (see Figure 8.9). This yields:

$$\frac{\tan\left(\Delta\phi_{\text{eff}}\right)}{\tan\left(\Delta\phi_{\text{px}}\right)} = \frac{f_{\text{sao}}}{kf_{\text{mao}}} \tag{8.32}$$

when dividing Equation 8.23 by Equation 8.22 and replacing the effective pitch difference by the fraction of the pixel pitch $\Delta p_{\text{eff}} = p_{\text{px}}/k$ from Table 8.2. The focal length of the single-aperture imaging optics is denoted by f_{sao} and that of a single channel of the multiaperture imaging optics by f_{mao}. The variable k is an integer number according to Table 8.2. The same angular sampling has to be achieved for the multiaperture system compared to the single-aperture system, as this is desired for the application. Hence, the condition yields

$$\frac{f_{\text{sao}}}{kf_{\text{mao}}} \overset{!}{=} 1 \quad \rightarrow \quad f_{\text{mao}} = \frac{f_{\text{sao}}}{k}. \tag{8.33}$$

In conclusion, the effective focal length of the multiaperture optical system can be shortened by $1/k$ while the angular sampling and the pixel size are constant. The integer $k \geq 1$ is called the *braiding factor*.

TABLE 8.2 Summary of the Different Types of Effective Sampling for the Multiaperture Imaging Systems Using Electronic Stitching of Segments from Figures 8.8 through 8.10

Parameter	(1) Redundant Sampling	(2) Increased Sampling	(3) Simple Stitching
Pitch difference	$\Delta p_K = p_{\text{px}}/m_0$ m_0 is integer >1	$\Delta p_K = l_p/k$ k is not a divisor of n_g	$\Delta p_K = l_p$
Effective sampling in image space	$\Delta p_{\text{eff}} = p_{\text{px}}/m_0$	$\Delta p_{\text{eff}} = p_{\text{px}}/k$	$\Delta p_{\text{eff}} = p_{\text{px}}$
Effective sampling in object space	$\Delta\varphi_{\text{eff}} = \Delta\varphi \approx \Delta\varphi_{\text{px}}/m_0$	$\Delta\varphi_{\text{eff}} \approx \Delta\varphi_{\text{px}}/k$	$\Delta\varphi_{\text{eff}} = \Delta\varphi_{\text{px}}$

8.6.2.2 Examples of Systems Using Electronic Stitching of Segments

Redundant Sampling: The ANSE enables color imaging and an increase of the SNR by a factor of 3 compared to APCOs by applying a redundant sampling in the FOV [56]. They belong to the thinnest artificial compound eyes with a thickness of <450 μm (see Figure 8.11), but they suffer from a low image resolution of 60 × 43 pixels and small information capacity (Figure 8.26). Nevertheless, such a setup is suitable for applications of ultrathin sensors with unlimited depth of field.

Increased Sampling: The APCO offers the shortest total track length by applying a single pixel per optical channel (which means a minimum possible number of pixels per channel). On the other hand, a single-aperture imaging system combines all pixels in a single optical channel (which is the maximum number of pixels per channel) with a comparably large total track length. The gap

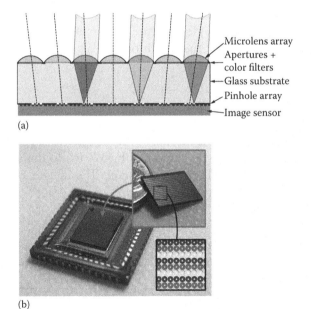

(a)

(b)

FIGURE 8.11 (a) Layout of the artificial neural superposition eye (ANSE) with integrated polymer color filters to acquire color images. The case of $n_g = 3$ is shown here. Each point in object space is imaged through a red, green, and blue color filter in different channels. (b) ANSE optics module for color imaging (box top right) directly attached to a CMOS image sensor (type Saentis, ZMD). The microlens array with integrated color filters is shown in the inset (box bottom right). (From Brückner, A., Duparré, J., Dannberg, P., Bräuer, A., and Tünnermann, A., *Optical Express*, 15, 11922–11933, 2007.)

in between these extremes is filled by the multiaperture imaging systems, which use electronic stitching of segments. Here, a partial image of a certain number of pixels is captured in each channel. The partial images are then flipped and stitched together by means of software processing, hence the name electronic cluster eye (eCLEY). The increased sampling that is introduced by the braided FOVs of adjacent optical channels is advantageous for achieving a short total track length compared to a single-aperture optical imaging system (see Section 8.5). In the practical realization, a braiding factor of $k = 2$ was chosen so that the effective pitch difference is

$$\Delta p_{\text{eff}} = \frac{p_{\text{px}}}{2} \quad \text{and} \quad \Delta p_{\text{K}} = \frac{n_g p_{\text{px}}}{2}. \tag{8.34}$$

Thus, the sampling in object space is effectively doubled and the total track length can be reduced to about half that of comparable single-aperture optics.

8.6.2.3 Optical Design for Thin Wafer-Level Camera Optics

Starting with the analytical relationships from Section 8.4, an optical design study was undertaken in order to explore the possible solution space. Four different configurations were selected for the optimization and simulation using ray-tracing methods. Table 8.3 lists their basic properties. Two of the described optical designs have been realized, namely, the eCLEY 200p and the eCLEY VGA. In terms of image resolution, the eCLEY 200p is the counterpart of the former APCO but with the advantage of a three times smaller lateral footprint size (Figure 8.12a). Due to the larger number of pixels per channel, the information capacity of the eCLEY 200p is 200% higher than that of the APCO and the 2D optical fill factor of an eCLEY is generally up to two orders of magnitude higher (compare 0.34% from Table 8.1 to the values in Table 8.3). The eCLEY VGA was designed with the specifications of a wide-angle camera module for VGA resolution (640 × 480 pixels). In contrast to placing the aperture stop at the lens (as in the eCLEY 200p), a new degree of freedom was necessary to achieve the higher image resolution. Thus, the stop was moved to a position in front of the lens where the coma for an off-axis field is nearly completely corrected. A tuning ("chirp") of the tangential and sagittal radii of curvature of each microlens according to its individual viewing direction was carried out

TABLE 8.3 Investigated Optical Designs for eCLEYs of Different Complexity and Resolution

Demonstrator	eCLEY 200p	eCLEY VGA	eCLEY 1MP	eCLEY 720p
Diagonal FOV (degree)	65	70	64	70
F-number	2.4	3.7	2.8	2.8
Image resolution (pixels)	200×200	660×500	1200×840	1440×864
Number of channels	23×23	17×13	13×9	15×9
Number of pixels/channel	9×9	39×39	93×93	96×96
Pixel pitch (μm)	3.2	3.2	2.2	2.2
Total track length (mm)	0.47	1.42	1.67	1.92
Footprint size (mm²)	3.5×3.5	6.8×5.2	11.5×8.0	8.0×4.9
Lenslet diameter[a] (mm)	0.131	0.375	0.840	0.514
FOV per channel (degree)	5.3	7.8	9.0	9.2
2D optical fill factor (%)	5	14	8	31
Layout				
Microlenses/channel	1	1	3	2
Microlens shape	Toroidal	Toroidal	Toroidal	Free-form, polynomial

[a] Value for the center microlens of the front microlens array.

in order to compensate for astigmatism and field curvature for the central viewing directions of all channels [39,41]. A material with a high refractive index (Exfine CO160, $n = 1.633$ at $\lambda = 588$ nm) was used for the lenslets in order to minimize residual field curvature across the limited field inside each channel. Additionally, the reversed MLA causes a higher telecentricity because ray bundles are refracted at the substrate front face before hitting the individual microlens and the substrate acts as a cover glass (Figure 8.12b).

In addition to the simulation of the optics MTF, spot diagrams, and distortion, a nonsequential ray-tracing model was created to optimize the suppression of optical cross talk. A fill factor $\Gamma \leq 0.4$ was chosen as a starting point. The axial position and the size of the openings of three additional diaphragm arrays were optimized for a sufficient suppression of cross talk.

(a)

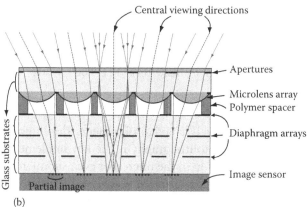

(b)

FIGURE 8.12 (a) Photograph of the eCLEY 200p optics chip attached to a CMOS image sensor. (b) Schematic cross section of the optical layout of the eCLEY VGA demonstrator. The optical module consists of two components: the microlens array with the aperture array and a spacer stack containing the diaphragm arrays. Both are fused by molded polymer spacers.

An additional aperture array on the front side of the MLA substrate absorbs light, which would otherwise hit the intermediate space between lenses, causing stray light. A detailed description of the simulation methods is found in Brückner [34]. An analysis of the simulated final image yields a moderate veiling glare index of 3.3%.

The nonsequential model was also used for testing the image stitching and the distortion correction algorithm from simulated images (Figure 8.13).

8.6.2.4 Fabrication of eCLEY with VGA Resolution

The fabrication of microoptical systems relies on the established processes and equipment of microtechnology [59–61]. Such a system is fabricated by a sequence of process steps that require

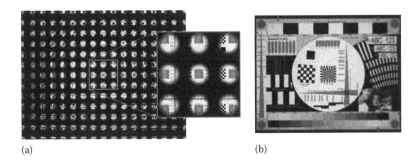

(a)   (b)

FIGURE 8.13 Simulated images of the eCLEY using a nonsequential ray-tracing method and a two-dimensional representation with Lambertian illumination for the target object. (a) The simulated image of a CCTV test chart as captured from the image sensor. The inset shows a magnified section. (b) The final image after applying the stitching algorithm. The right half of the image includes the simulation of diffraction blur. The final image resolution is 660 × 500 pixels. About 12 billion rays were traced to obtain the result.

precise lateral as well as axial alignments throughout the whole process chain. Photolithography is the key implement to achieve the required precision in the micron or even submicron range. The highly precise lateral alignment is achieved for many components in parallel using alignment marks on a planar mechanical carrier substrate called *wafer* (Figure 8.14).

The fabrication of master structures and replication tools was carried out with photolithography and reflow according to the description in Figure 8.14. However, the two main components of the system were processed separately: the MLA wafer, which makes up the front part of the stack, and the spacer wafer, which is a laminate of three glass substrates. The diaphragm arrays were structured on both sides of the first glass wafer with UV lithography using a black-matrix polymer resist (PSK 2000, Brewer Science). A second, unstructured wafer was bonded after coating the structured wafer with a thin adhesive layer and another diaphragm array was structured on the plain side. Subsequently, the lamination was repeated for a third glass wafer forming a spacer wafer with a total thickness of 1.1 mm. Meanwhile, the MLAs and polymer spacers of 20 μm height were molded in two steps on a second preprocessed glass wafer. Finally, the microlens wafer was integrated upside down with the glass spacer substrates at wafer level. The shallow polymer spacers on top of the MLAs cause the necessary air gap between the two components. Such a chain of vertical integration of binary

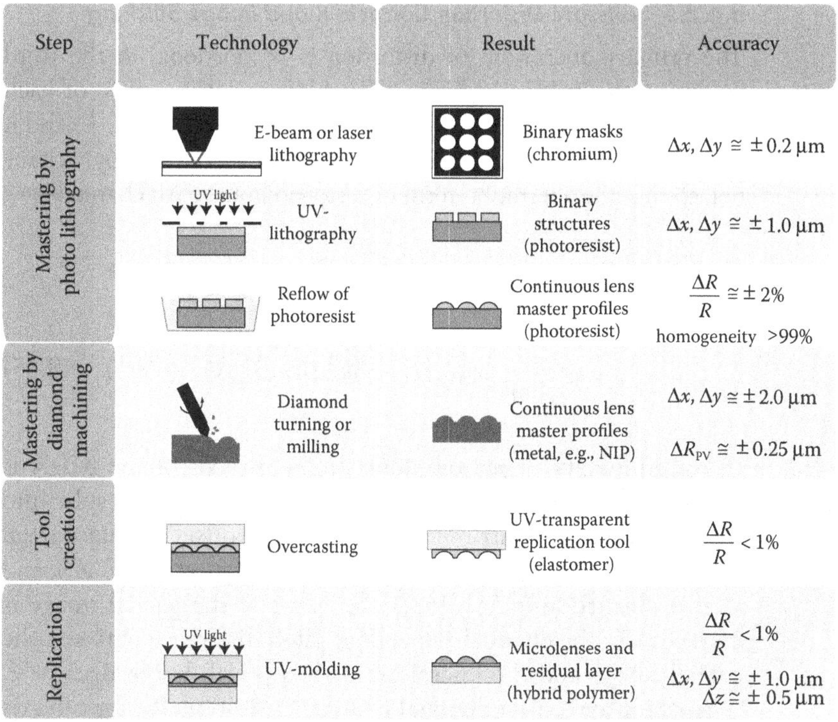

Step	Technology		Result		Accuracy
Mastering by photo lithography	E-beam or laser lithography		Binary masks (chromium)		$\Delta x, \Delta y \cong \pm 0.2\ \mu m$
	UV light UV-lithography		Binary structures (photoresist)		$\Delta x, \Delta y \cong \pm 1.0\ \mu m$
	Reflow of photoresist		Continuous lens master profiles (photoresist)		$\dfrac{\Delta R}{R} \cong \pm 2\%$ homogeneity >99%
Mastering by diamond machining	Diamond turning or milling		Continuous lens master profiles (metal, e.g., NIP)		$\Delta x, \Delta y \cong \pm 2.0\ \mu m$ $\Delta R_{PV} \cong \pm 0.25\ \mu m$
Tool creation	Overcasting		UV-transparent replication tool (elastomer)		$\dfrac{\Delta R}{R} < 1\%$
Replication	UV light UV-molding		Microlenses and residual layer (hybrid polymer)		$\dfrac{\Delta R}{R} < 1\%$ $\Delta x, \Delta y \cong \pm 1.0\ \mu m$ $\Delta z \cong \pm 0.5\ \mu m$

FIGURE 8.14 Overview of the basic steps for the fabrication of microoptical multiaperture imaging optics. The illustrated steps are feasible at wafer level with the indicated accuracy.

and continuous microstructures is crucial for the fabrication of high-resolution microobjectives at the wafer level. The resulting optics wafer stack was diced (see Figure 8.15a) and single optics chips were aligned to the 3 MP image sensors (Micron MT9T001, Figure 8.15b) [62].

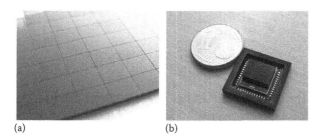

(a)   (b)

FIGURE 8.15 (a) Diced wafer with several optics modules of the eCLEY VGA. (b) Size comparison between the eCLEY VGA assembled on an image sensor array and a one-cent coin. (From Brückner, A., Duparré, J., Leitel, R., Dannberg, P., Bräuer, A., and Tünnermann, A., *Optical Express*, 18, 24379–24394, 2010.)

8.6.2.5 Software Distortion Correction and Image Stitching

The primary aberration of distortion is proportional to the third power of the field coordinate [27]. However, the center of each channel of the eCLEY is directed at a chosen field point, h_i, and it varies only by a small amount, Δh, within the FOV of a single channel, so that the aberration term of distortion of the ith channel is

$$\delta^i_{\text{distortion}} = \frac{R}{n} \cdot a_{311} \cdot \left(h_i + \Delta h \right)^3$$

$$= \frac{R}{n} \cdot a_{311} \cdot \left(h_i^3 + 3h_i^2 \Delta h + 3h_i \Delta h^2 + \Delta h^3 \right). \quad (8.35)$$

For simplicity, a section along one main axis of the MLA is assumed. The distortion of the ith optical channel is split into the sum of four contributions: a constant offset of the partial image position ($\propto h_i^3$), a linear ($\propto h_i^2 \Delta h$), a quadratic ($\propto h_i \Delta h^2$), and a cubic distortion ($\propto \Delta h^3$) [63]. The offset of the partial image is corrected by the channel-dependent pitch difference between the partial image and the microlens in the optical design because h_i is constant for each individual channel. Additionally, for off-axis channels the condition $\Delta h \ll h_i$ is fulfilled and the linear term ($\propto \Delta h$) in Equation 8.35 is the dominating part of the distortion of the individual channel.

Partial image distortion has to be corrected in order to yield a good stitching of the image details from the braided sampling of the scene by the eCLEY. A bilinear spatial transformation is used for the software distortion correction because for the aforementioned reason it is a good approximation and it is simple and fast.

$$\tilde{x} = c_1 x + c_2 y + c_3 xy + c_4 \qquad (8.36)$$

$$\tilde{y} = c_5 x + c_6 y + c_7 xy + c_8 \qquad (8.37)$$

The variables x, y and $\tilde{x}$, $\tilde{y}$ denote the undistorted and distorted coordinates, respectively. The coefficients c_i of the transform can be calculated from solving the equation system for four different reference points. Once the coefficients are known, the inverse transform is able to create an undistorted image from the distorted input image. In practice, the eight coefficients are tabularized for each optical channel. Furthermore, an interpolation step is necessary after the spatial transform because the equidistant pixel grid of the captured image is transformed into a nonequidistant grid. The exactness of the correction depends on the accuracy of the model in relation to the physical distortion and the method of interpolation. If the calculation is carried out for each color plane individually, the correction of the lateral chromatic aberration is feasible by the same procedure as it is linearly dependent on the field coordinate.

The initial image-processing pipeline contained: (1) Bayer demosaicing and color correction done by the sensor readout software, (2) extraction of the partial images from the full image matrix (Figure 8.16a), (3) channelwise bilinear correction of distortion, (4) channelwise linear interpolation of undistorted partial image pixels on a regular sampling grid, and (5) mapping of the undistorted pixels of each partial image into the final image matrix including horizontal and vertical flipping as well as braiding. The last step uses the deterministic algorithm of increased (braided) sampling in order to resemble the original neighborhood of each pixel in object space. An example of the result is shown in Figure 8.16b.

8.6.2.6 Experimental Characterization of eCLEY VGA

The processing time for the distortion correction, interpolation, and image stitching was measured to be 20 ms (Intel CoreDuo machine 2.66 GHz), which makes it feasible for real-time capture and display. Processing and capturing frames can be interlaced so that frame rates of >30 fps are possible on standard hardware.

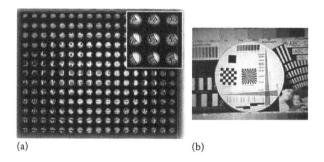

(a)   (b)

FIGURE 8.16 Image of a CCTV test chart that was captured with the prototype of the eCLEY VGA. (a) Image as recorded by the image sensor with a full resolution of 2050 × 1540 pixels. The inset shows a magnified section of the 17 × 13 partial images. (b) Final image (660 × 500 pixels) after distortion correction and image stitching. (From Brückner, A., Duparré, J., Leitel, R., Dannberg, P., Bräuer, A., and Tünnermann, A., *Optical Express*, 18, 24379–24394, 2010.)

Some defects such as edge displacements ("zipper artifacts") and color aliasing were found when examining the image quality (see star in the center of Figure 8.16b). However, they are very difficult to quantify. They arise from irregular sampling over the object field, which is due to a combination of distortion correction and parallax between adjacent channels [64]. Recent enhancements of the distortion and parallax correction algorithm have led to an improved stitching without zipper artifacts and the possibility for software focusing without moving parts [64,65].

The suppression of optical cross talk was tested using a collimated HeNe laser source in front of the eCLEY prototype, which was mounted on a goniometer in order to change the angle of incidence for the illumination. The first ghost images appeared for large integration times so that a suppression factor between the ghost image and the signal of about 1:20,000 was determined. The difference between the measured value and the simulation results from the fact that the simulated light source in the image plane emitted rays into the half space, whereas the real image sensor pixels have a finite angular sensitivity. Thus, the simulation is a worst-case scenario. Additionally, the eCLEY was used for image capture outside on a bright sunny day (see Figure 8.17a), which is one of the most challenging scenarios for stray light and cross talk and it is close to practical use.

8.6.2.6.1 MTF Performance

The MTF of the eCLEY VGA was measured using a slanted edge target (according to ISO 12233) [66]. The results for two field positions are shown in Figure 8.18. The simulated MTF curves

(a) (b) (c)

FIGURE 8.17 Images captured with the eCLEY VGA prototype. (a) Snapshot of the Fraunhofer Institute IOF in Jena. (From Brückner, A., Duparré, J., Leitel, R., Dannberg, P., Bräuer, A., and Tünnermann, A., *Optical Express*, 18, 24379–24394, 2010.) (b) Slanted edge chart as used for the MTF measurements. (c) Image of a regular grid for measuring distortion.

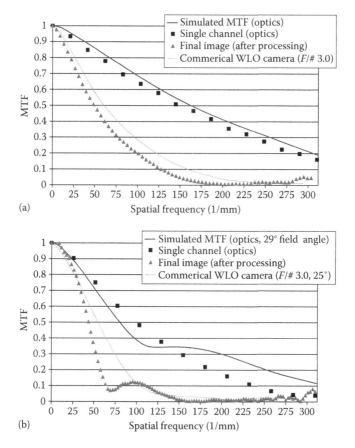

FIGURE 8.18 (a) Comparison between different polychromatic on-axis MTF curves for the eCLEY VGA. (From Brückner, A., Duparré, J., Leitel, R., Dannberg, P., Bräuer, A., and Tünnermann, A., *Optical Express*, 18, 24379–24394, 2010.) (b) The same for an off-axis field with an angle of incidence of 29°. Solid line: Simulated optics MTF for the specific angle of incidence. Squares: MTF of the optics measured without an image sensor. Triangles: MTF measured in the final image of the fully assembled prototype (optics and image sensor) including image post-processing. Thin gray line: MTF measured from a commercial camera module using the same method. The object distance was 25 cm for all measurements.

(shown as black solid lines) were derived from ray-tracing software for a single object field coordinate using a calculation method that accounts for diffraction. In order to measure solely the optics MTF (illustrated by black squares), the optical module was placed in the center of a rotation table and the partial image plane of a selected channel was relayed on a charge-coupled device (CCD) camera chip by a 40× microscope objective. Hence, the partial image of the edge object that was created by a single optical channel of the eCLEY was captured with high magnification. The third MTF curve (triangles) was measured in the final image of the assembled prototype with the optics mounted to the complementary metal-oxide-semiconductor (CMOS) image sensor. It includes the influence of the whole image-processing chain.

The Nyquist frequency of the system is 312.5 cycles/mm because the effective pixel pitch in the final image is half the sensor pixel pitch (3.2 µm) due to the braiding factor $k = 2$. The measured single channel optics MTF resembles the simulated MTF well. There is a systematic offset of <10% which is caused by fabrication tolerances and residual defocus due to an incorrect spacer thickness. A significant reduction of the total MTF (by up to 30%) is caused during the image acquisition by the image sensor. The reason is the aperture MTF of the image sensor pixels as well as a further reduction of the spatial resolution due to pixel cross talk and the effects of the Bayer CFA (Figure 8.19a).

It has to be noted that the optical fill factor of the image sensor pixel is not exactly known and thus its influence on the MTF is uncertain. The MTF performance is further decreased by about 10%–15% due to the partial image distortion correction and stitching algorithm (Figure 8.19a). Although the fabricated optics module has the ability to transfer spatial frequencies higher than the system Nyquist frequency, the MTF of the final image vanishes at about the image sensor Nyquist frequency of 156 cycles/mm in Figure 8.18.

Another MTF curve (gray line) was added in Figure 8.18 for comparison with a commercial single-aperture WLO camera module with a VGA resolution and a total track length of 2.2 mm (OmniVision OVM7690 CameraCube [67]). It yields a 10% advance at midspatial frequencies over the MTF of the final (braided) image of the eCLEY VGA. However, the images of the WLO camera contain more noise due to the small pixel size of 1.75 µm.

8.6.2.6.2 Distortion and Relative Illumination

The distortion was measured in an image of a regular grid (see Figure 8.19b) which has knots at the points where the optical axes of the individual channels meet the object plane at a given object distance. The correction of interchannel distortion was verified by measuring the image coordinates of the knots in the unprocessed image from the image sensor. An average distance

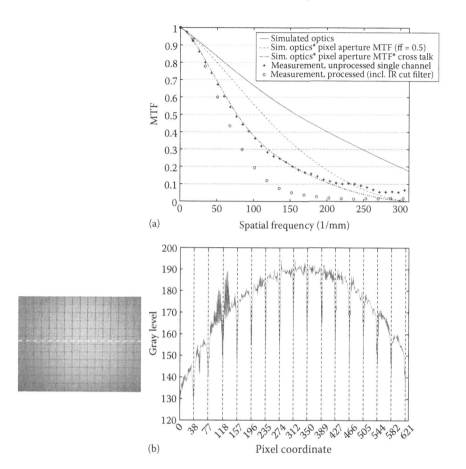

(a)

(b)

FIGURE 8.19 (a) Measured influence of the software image stitching on MTF. All curves show the polychromatic MTF in the center of the field for about 50 cm object distance. The measured MTF in a single channel (crosses, prior to post-processing) can be reproduced by a simulation that considers the pixel aperture MTF (dashed line, at a fill factor of 0.5) and pixel cross talk (dash-dotted line). The MTF in the final image (circles) is decreased due to the image stitching post-processing. (b) Measurement of distortion from a line scan through the grayscale image of a regular grid. The scan path is indicated by the dashed line. The measured center of each grid line is marked by a tick label on the pixel coordinate axis. (From Brückner, A., Duparré, J., Leitel, R., Dannberg, P., Bräuer, A., and Tünnermann, A., *Optical Express*, 18, 24379–24394, 2010.)

of 111 pixels was measured which equals the pitch of the partial images $p_K = 355.2$ µm. Thus, there is no interchannel distortion within the accuracy of the measurement.

Additionally, the distance of each pair of grid lines was measured across the FOV in the final braided image to determine the residual distortion of the partial images (Figure 8.19b). A mean distance between adjacent lines of 38.87 pixels was measured and compared to the designed value (39 pixels), which gives a deviation of 0.3%. No monotonic distribution of distortion was found across the final image. Thus, the image of the eCLEY VGA is nearly free of distortion. However, insufficiencies of the software correction are difficult to quantify from the final images of the eCLEY because displaced image details (zipper artifacts) might also be caused by irregular sampling due to parallax or residual misalignment between the optics module and the image sensor.

The relative illumination in image space was also analyzed from Figure 8.19b. The irradiance in the image decreases to about 69% across a horizontal scan, which corresponds to a full field angle of about 50°. This is in good agreement with the simulation. An overview of the measured parameters of the eCLEY VGA demonstrator is given in Table 8.4.

8.6.2.7 Improvements of eCLEY

The degrees of freedom in the optical design of a single channel have to be increased in order to create eCLEY optics of a higher resolution and a lower *f*-number. For example, a triplet design

TABLE 8.4 Summary of Simulated and Measured System Parameters of the eCLEY VGA

Parameter	Simulation	Experiment
Diagonal field of view (°)	70	70.1 ± 0.35
Radius of curvature[a] (μm)	492–531	493–533
RMS surface deviation (nm)	–	55
Irradiance difference between 0° and 25° field angle (%)	14.5	19.4
Optics MTF,[b] on-axis, at $v_{Ny}/4 = 78$ cycles/mm (%)	75	71
Optics MTF,[b] on-axis, at $v_{Ny}/2 = 156$ cycles/mm (%)	51	48
Capture MTF,[c] on-axis, at $v_{Ny}/4 = 78$ cycles/mm (%)	66.5	38
Capture MTF,[c] on-axis, at $v_{Ny}/2 = 156$ cycles/mm (%)	32	5
Spatial resolution limit (capture MTF = 10%) (mm⁻¹)	233	125
Number of resolvable image points	660 × 500	530 × 406
Average image distortion (%)	–	0.3

[a] Values denote minimum and maximum data of the chirped MLA.
[b] Simulation and measured data from the solid curve and squares in Figure 8.18a, respectively.
[c] Simulation data from the dashed curve and measured from blue circles in Figure 8.19a. The system Nyquist frequency is $v_{Ny} = 312.5$ cycles/mm due to the increased sampling with a braiding factor of $k = 2$.

with three MLAs was investigated for the eCLEY 1 MP. The quasi-symmetrical arrangement with respect to the aperture stop corrects most of the coma, distortion, and lateral color aberrations. However, the limitation of the chirped arrays of spherical microlenses leads to a laterally enlarged system with a low optical fill factor due to the large diameters of the individual lenslets (see Table 8.3). The correction of spherical aberrations and field curvature in each channel is also limited, especially for off-axis field coordinates, when using spherical lenslets at low *f*-numbers (e.g., 2.8). Another step of complexity takes advantage of the fact that only a segment of each lenslet is needed to focus the respective parts of the FOV. The 2D fill factor is increased by about a factor of 4 if the lens segments are cut and packed close to each other (Figure 8.20a). This leads to a complex surface structure with steep slopes between off-axis lens segments, which cannot be mastered by reflow of the photoresist. Hence, a pointwise mastering technology has to be used so that the lens surfaces may have arbitrary profiles in order to yield good aberration corrections. This complex microstructure is further called *refractive free-form array* (RFFA). The eCLEY 720p was designed to include two RFFA components, which is one optical surface less in comparison with the eCLEY 1 MP, but with the advantage of an improved fill factor (from 8% to 31%) and an increased resolution (Figure 8.20b). The surface profile of the individual lenslets (*z*) is described using a polynomial in the local coordinates of the aperture plane *x*, *y*:

$$z = \frac{c_v\left(x^2 + y^2\right)}{1 + \sqrt{1 - \left(1 + k_c\right)c_v^2\left(x^2 + y^2\right)}} + \sum_{i=1}^{N} C_i P_i\left(x, y\right) \quad (8.38)$$

where
  c_v is the curvature
  k_c is the conic constant
  C_i are the polynomial coefficients
  P_i are the N polynomial terms

Twenty-five polynomial terms were applied for the eCLEY 720p ranging from the second to the sixth order in *x*, *y*. The optimization of the surface profile of each channel as well as the simulation and an image analysis were carried out by ray

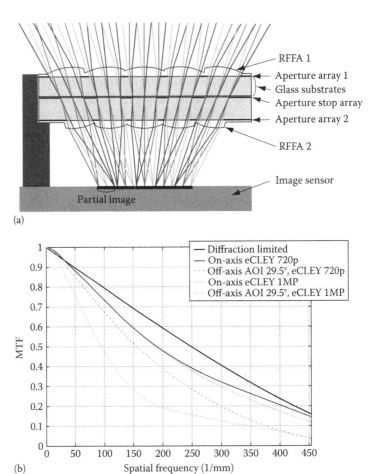

(a)

(b)

FIGURE 8.20 (a) Schematic section of the eCLEY 720p which consists of a stack of two glass substrates with refractive free-form arrays (RFFA) on either side. (b) Simulated polychromatic MTF curves of the eCLEY 720p for different field positions. When compared to the curves of the eCLEY 1MP, the RFFAs demonstrate much better off-axis MTF values (e.g., 20% increase at 200 cycles/mm). The angle of incidence (AOI) for both off-axis fields is 29.5°.

tracing according to the methods described by Brückner [34]. The optimized RFFAs were exported into a CAD file format (Figure 8.21c) in order to fabricate a master mold by ultraprecision diamond machining. Figure 8.21a,b demonstrates the result of the image simulation for the current optical design.

Unfortunately, the amount of optical cross talk is increased by the large aperture size of the lenslets and the enhanced fill factor of the pixel groups in the image plane. An additional glass substrate with two diaphragm layers was added near the image plane

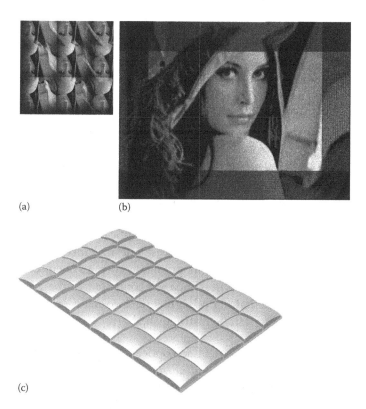

(a) (b)

(c)

FIGURE 8.21 Image simulation for an eCLEY with RFFA elements: (a) About 3 × 3 channels of the full array of partial images after tracing 1 billion rays. (b) Simulated final image after software stitching. Only a quarter of the full array was simulated in order to reduce the tracing time. The image size is 740 × 560 pixels. (c) Layout rendering of the front RFFA of eCLEY 720p. Only a quarter of the array is shown.

in the current optical design to suppress cross talk. The nonsequential simulations yielded the strongest ghost of 9% of the useful signal intensity and a veiling glare index of 6.9%. Thus, future work will investigate the fabrication of vertical separation walls, for example, by two-photon lithography, in order to provide sufficient suppression of cross talk for systems with RFFAs.

Simple Stitching: Recently, an artificial compound eye has been proposed which is representative of the simple stitching approach in that a part of the FOV is recorded in each individual channel and image processing is used to fuse the different partial images to an overall image of the whole FOV [14]. Its purpose is to create a compact imaging system called *MULTICAM* for application in the infrared spectral range [14,68]. For the sake of a high optical fill factor and ghost light suppression, a stack of several microlenses is used. Additionally, a prism array is applied to increase the FOV to 30°. This compromises the compactness of the device (total track length is 24.8 mm) and its light sensitivity ($F/\# = 8$).

8.6.3 Optical Stitching of Segments

The optical stitching of segments denotes a multiaperture imaging system that achieves a seamless image of the total FOV by an optical recombination of all partial images (see Figure 8.6 *middle left*). In this case, a conventional image sensor is applicable and the optical fill factor yields $\Gamma^2 = 1$. The lateral dimensions of such a multiaperture objective are comparable to those of single-aperture systems. However, as the optical stitching is achieved by an imaging setup with at least one intermediate image, the number of optical elements per channel, the system complexity, and the overall thickness increase compared to the approach of the electronic stitching of segments. Two main principles of the optical stitching can be distinguished:

> *Side-by-Side Stitching*: Each optical channel transfers its related part of the FOV and adjacent partial images are optically stitched together in such a way that the image details at the intersections between adjacent partial images are preserved (Figure 8.22a). Side-by-side stitching yields an artificial compound eye of high resolution that is especially compact

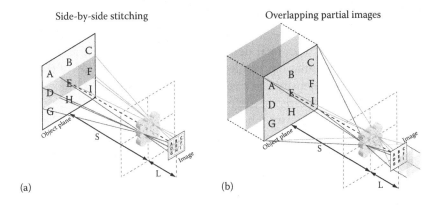

(a)   (b)

FIGURE 8.22 Overview of different object sampling solutions for the optical stitching of segments. (a) The side-by-side stitching as used by the optical cluster eye (oCLEY). (b) Individual partial images are superimposed in the image plane. This scheme is found in the Gabor superlens (GSL). The object distance and the thickness of the optical system are denoted by s and L, respectively.

in the lateral dimensions. The principle is used by the *optical cluster eye* (oCLEY) [35,17,69] and the *ultrathin array microscope* [70,71].

Overlapping Partial Images: Each optical channel images a larger portion of the full FOV and the individual partial images overlap in the image plane (Figure 8.22b). Hence, the light collection ability of the artificial compound eye can be considerably increased. This is the working principle of the *Gabor superlens* (GSL), which is the technical counterpart of the natural superposition compound eye [1].

Due to the working principle, there is no possibility to increase the sampling in object space beyond that given by the pixel pitch in Equation 8.22.

8.6.3.1 Examples of Systems Using Optical Stitching of Segments

Side-by-Side Stitching: The oCLEY captures a wide FOV on a conventional, densely packed image area. It belongs to the class of multiaperture optics that applies optical image stitching and thus overcomes a major drawback of the principle of electronic stitching of segments, namely, the need for a customized image sensor of a larger active area. The first prototypes of such a system were demonstrated in 2004 [17,35]. However, the first solutions did not achieve a satisfactory image stitching and focusing performance [17]. The following work led to a prototype of the oCLEY shown in Figure 8.23, which achieved an information capacity of 446×335 resolvable image points on an extremely small sensor footprint size of 1.44×1.08 mm [69]. The small lateral footprint yields a benefit for the chip costs. With a total track length of 1.86 mm, the array of microobjectives with tilted optical axes is shorter than single-aperture objectives with a comparable resolution and information capacity (Figure 8.26).

In contrast to prior implementations, a focused image was achieved with optical partial image stitching of subpixel accuracy by implementing an additional MLA that tilts the optical axes of the off-axis channels toward normal incidence. The strong change of relative illumination across the image could be improved by applying an image sensor with an adapted fill factor enhancing the MLA. However, with an *f*-number of $F/\# = 6.7$, the light collection ability is 3.3 times lower than that for the eCLEY as the ratio of the irradiance in the image plane of two systems 1 and 2 scales according to $E_{im,1}/E_{im,2} = (F/\#_2/F/\#_1)^2$ [15,23,25,33]. The

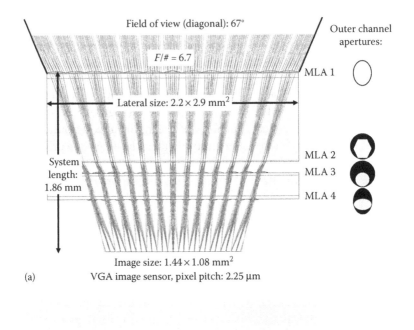

(a)

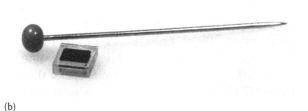

(b)

FIGURE 8.23 (a) Overview of the final optical design of the oCLEY demonstrator with VGA resolution. (b) Comparison of one of the fabricated and diced optical modules to a pin. (From Meyer, J., Brückner, A., Leitel, R., Dannberg, P., Bräuer, A., and Tünnermann, A., *Optical Express*, 19, 17506–17519, 2011.)

tradeoff between *F/#* and the system length, which is inherent to the oCLEY principle, makes it unlikely to be improved without increasing the system length. Thus, the oCLEY is not able to achieve an *f*-number lower than 4.5 at a competitive total track length for miniature camera applications [34]. However, it yields a compact optical imaging system with a large depth of field that can be manufactured at wafer level.

Close-up Side-by-Side Stitching: Arrays of gradient-index (GRIN) lenses in a two or three row configuration such as that found in photocopying machines and flatbed scanners can be regarded as a special case of a side-by-side optical image stitching with unity magnification [72,73]. Each lens forms an erect image, which superposes with the images of neighboring channels. This is achieved by a symmetric setup of lens arrays without pitch difference. The 2D image is generated by scanning the lines perpendicular to the orientation of the array. A similar optics is found in contact image sensor (CIS) modules that are widely used in industrial inspection [74]. Either the imaging system or the object has to be moved, which causes a rather large setup with mechanically limited resolution. Also nonscanning, multiaperture optics with unity magnification have been reported [75], which principally overcome the scanning limitations by introducing 2D lens arrays for optical recording on photographic film [76] or with application in projection mask lithography [77,78]. Lately, this principle has been applied to obtain an ultracompact digital imaging microscope with unity magnification by stacking symmetrical MLAs of equal pitch [70,71]. Such a setup decouples the tradeoff between the system length and the size of the object field. Hence, the captured object field can be enlarged by simply increasing the number of optical channels. A prototype of the ultrathin array microscope was integrated on a full-format image sensor so that a field size of 36.1×24.0 mm was captured by a system as short as 4.0 mm in a single exposure (see Figure 8.24) [71]. The experimental resolution limit was measured to be 83 cycles/mm (or 12 μm structures) across the object field, taking into account the residual imperfections of the partial image stitching. When looking at the simulation results, it is promising that the resolution could be increased up to 200 cycles/mm with tight control of fabrication tolerances and sophisticated assembly techniques.

An improved version with integrated front-side illumination was realized by adding a flat light guide with an additional microstructure for controlled light out-coupling [79]. The increase of the numerical aperture for low-light tasks such as fluorescence imaging as well as the improvement of the stability of the fabrication process are the key tasks for future work in order to enable the array microscope to enter applications in industrial inspection, quality assessment, and biomedical imaging.

Overlapping of Partial Images: In 1940, Dennis Gabor [80] published the basic scheme of an artificial compound eye using the overlap of partial images. A related patent claims a stack of two lens arrays with slightly different pitch axially separated by the sum of their focal lengths, which would act like a conventional

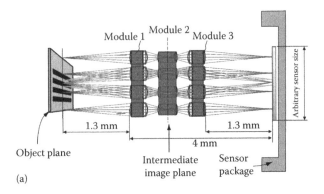

Module 1 Module 2 Module 3

1.3 mm

Object plane

1.3 mm

4 mm

Intermediate image plane

Sensor package

Arbitrary sensor size

(a)

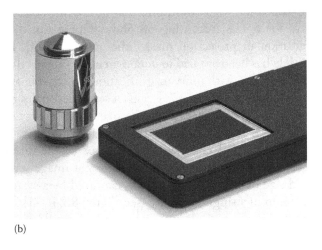

(b)

FIGURE 8.24 (a) Schematic layout section of the working principle of the ultrathin array microscope. (b) Photograph of the prototype's head in comparison with a conventional microscope objective lens.

lens of larger thickness (Figure 8.25a). The difference between the pitch of the first lens array, p_1, and the second, p_2,

$$\Delta p_{GSL} = p_1 - p_2 \tag{8.39}$$

leads to a back focal length, F_s, of the stack given by [81]

$$F_s = \frac{f_2 \cdot p_1}{\Delta p_{GSL}}. \tag{8.40}$$

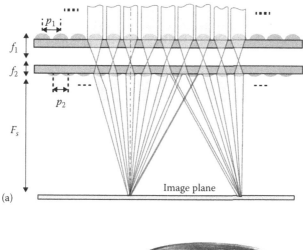

(b)

FIGURE 8.25 (a) Working principle of a Gabor superlens. (b) Size comparison of a fully assembled microoptical Gabor superlens (μoGSLs). (From Stollberg, K., Brückner, A., Duparré, J., Dannberg, P., Bräuer, A., and Tünnermann, A., *Optical Express*, 17, 15747–15759, 2009.)

Even if the focal length of the second lens array, f_2, is small, the focal length of a larger lens can be created by choosing the pitch difference, Δp_{GSL}, accordingly. Such a setup became known as the *Gabor superlens* when it was experimentally demonstrated for the first time, more than 50 years after Gabor's patent [81,82]. However, with a total track length of about 65 mm, a FOV of 10°, and a resolution of 2 line pairs/mm, this example was neither small nor suitable for an application.

The overlap of the partial images in the microoptical Gabor superlens (μoGSL) enables the realization of a high light collection ability while maintaining a compact size and thus overcomes the main drawback of the oCLEY. The first microoptical prototype yielded an *f*-number of 2.8 at a total track length of 2 mm (Figure 8.25b) [83]. This is 5.7 times more sensitive than the oCLEY. However, a large part of the overall FOV is transmitted by each optical channel, which causes additional blur across the single partial images due to off-axis aberrations and leads to

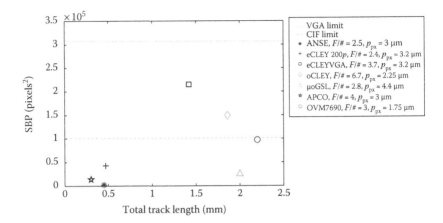

FIGURE 8.26 Experimentally derived space-bandwidth product (SBP or number of resolvable image points) of several realized MAOs as a function of the total track length. The smallest commercially available SAO system OVM7690 is added for comparison. (Adapted from OmniVision, OVM7690 640 × 480 CameraCube™ Device, Product brief, 1st ed., 2010.)

a comparably low spatial resolution and, thus, small information capacity (Figure 8.26). The FOV is limited to about 30° due to vignetting of off-axis rays. But, even this size of FOV and the small footprint size of 2.8 mm² provide a considerable improvement over former realizations [81]. Future work will have to investigate the extension of the FOV size and the increase of the information capacity.

8.6.4 Multiaperture Super-Resolution

The fourth approach to multiaperture imaging systems (Figure 8.6 *outer right*) has no known archetype in nature but it uses ideas of video image processing. The term *super-resolution* stands for the processing of a number of undersampled images with low resolution, for example, out of a video stream, in order to create a single image with high resolution [84]. Instead of capturing multiple images over time, they could be captured at the same time in multiple optical channels. Correspondingly, a low-resolution image of the full FOV is captured in each channel and an image registration and fusion algorithm is used to create the final, high-resolution image from all subimages. It has been proposed that this approach is promising as a thin and inexpensive substitute for imaging optics in the infrared spectral range [13]. A Japanese research group first published the so-called TOMBO system [12], which used diffractive lenslet arrays, GRIN lenses, or conventional refractive MLAs to construct a thin camera device based on super-resolution in the visible spectral range [12,85]. It has been demonstrated that this device is suitable for color image acquisition and multispectral imaging by using red/green/blue (RGB) or several narrowband interference filters [86,87]. Additionally, it has been shown that 3D imaging of objects at a distance of <1 m is feasible [88].

However, multiaperture super-resolution has several limitations. First, the FOV of the published examples is below 22° due to the limited usable image field per channel, which is determined by the diameter of the related lens and the technological limits for the numerical aperture of the microoptics [12,89].

Additionally, the approach is sensitive to object distance due to a change of parallax and thus subpixel disparity between images of neighboring optical channels. Only small object distances of a few centimeters up to about 2 m are described in the literature. The subpixel disparity is also altered across the field by the distortion of each optical channel. The condition for fine-grained subpixel shifts between the different low-resolution images states the most crucial limitation: In contrast to a conventional image acquisition process, super-resolution requires aliasing of the low-resolution images within each optical channel. This can be expressed by a condition for the number of nonredundant channels, K_{ch} [90]:

$$K_{ch} = \frac{\nu_{px}}{\nu_{Ny}} = \frac{2}{\sqrt{\Phi_{IS}}}, \qquad (8.41)$$

where ν_{px} is the first zero of the pixel aperture MTF [91] and ν_{Ny} is the Nyquist frequency of the image sensor pixel pitch with $\nu_{Ny} = 1/(2p_{px})$. Both are important to consider in this case. The pixel fill factor, Φ_{IS}, is defined by the ratio of the photosensitive area of the photodiode, $A_{ps} = d^2$, to the pixel area, $A_{px} = p_{px}^2$:

$$\Phi_{IS} = \frac{A_{ps}}{A_{px}}. \qquad (8.42)$$

This means that the optical MTF and thus the resolution of the lens system has to be high enough in order to transfer sufficient frequency content above the Nyquist frequency up to $\nu_{px} = 1/d$. However, this becomes difficult to achieve in practice when considering the tolerances of fabrication and alignment during assembly, especially for miniature cameras with a small pixel pitch and thus a high Nyquist frequency. Furthermore, the computational load and the processing speed are still challenging for a real-time image acquisition [92,93].

8.7 Conclusions and Outlook

Throughout this chapter, it was shown that the inspiration from the working principles of natural compound eyes could be used for a successful miniaturization of man-made optical imaging systems. Thus, the archetypes of compound eyes were analyzed in order to overcome the scaling limits that apply to today's single-aperture cameras. Two main advantages were adopted for technical solutions in the present work: First, the segmentation of the FOV leading to a reduction of the imaging system's total track length and the adaptation of the optical parameters according to the viewing angle of the specific channel. This yields a simplification of the optical setup and a resolution near the diffraction limit. Second, the equality of scale makes it suitable to apply microoptical fabrication techniques for the creation of lenses at wafer level with relaxed tolerances compared to single-aperture systems. The application of processes and equipment from photolithography enables high precision in the range of the visible wavelength for a large number of optical modules in parallel at the wafer level. The aligned stacking of wafers is carried out for assembly purposes and an intrinsic housing is achieved for the different MLAs.

A novel classification of the different types of artificial compound eyes was introduced, which sorts the solutions according to their characteristics of FOV segmentation and sampling. The classes of multiaperture imaging systems with electronic and optical image stitching were introduced by example demonstration systems. Both achieved a larger SBP in each individual channel, which makes them superior to the APCO type.

For the electronic image stitching, different extended parts of the overall FOV are captured by individual optical channels. Each optical channel operates separately so that the partial images are created and recorded individually. Due to the inversion of each partial image, the spatial information is disordered in the image plane. The final image is created by digital image processing using a deterministic correlation between the sampled object points and their corresponding location in the image matrix.

A special type, termed the *electronic cluster eye* (eCLEY), applies an overlap of the FOVs of adjacent optical channels with a defined subpixel offset and a reduced focal length. Hence, the scaling limits are overcome and the total track length is decreased by 50% compared to single-aperture imaging systems, whereas the object sampling and the SBP are maintained.

To date, the demonstrated thickness of 1.4 mm makes the eCLEY VGA the thinnest optics module in its class of resolution. The smallest commercially available module has a thickness of 2.2 mm while using even smaller pixels [67]. The necessity of image post-processing led to a paradigm shift in the optical design in the present work. Distortion and lateral chromatic aberrations were left uncorrected in favor of a simplified optical setup and relaxed fabrication tolerances. Subsequent to capture, these errors were corrected by software processing for each partial image in real time. Thus, the eCLEY VGA captured a large FOV with a residual distortion of <1%. The mastering of the MLAs was carried out by photolithography and reflow of the photoresist. UV-molding was applied for replication. These microoptical technologies enabled a highly precise fabrication of hundreds of lenslet arrays in parallel on a single wafer, which is suitable for potential production in high volume. This makes it the most promising candidate for mini-camera applications and is the reason for the detailed discussion in this chapter. However, the optical fill factor of the presented demonstrator was low, due to the mastering of MLAs by melting the photoresist. Consequently, a large megapixel image sensor was applied in order to achieve a VGA resolution that is not competitive for a camera application. But, the study of a customized image sensor and an adapted optical design for refractive free-form arrays proved that the fill factor can be increased from 8% to 31% with room for further improvements in the future.

The second investigated class of multiaperture imaging systems creates a seamless image on the sensor even though different parts of the full FOV have been transferred by different optical channels. The optical stitching of segments is enabled by a two-step imaging which resembles an array of focused Keplerian microtelescopes: A first lens group creates a demagnified, inverted intermediate image, which is magnified on the image sensor by a second lens group in each channel. Erect partial images result, which are required for the optical image stitching. Hence, the setup becomes more complex and yields a larger total track length compared to the electronic image stitching.

In conclusion, the eCLEY and the ultrathin array microscope demonstrated the highest potential for the miniaturization of imaging systems. The assembly techniques of the array microscope will be improved in order to exploit diffraction-limited resolution, which is feasible with today's microoptics technology. The integration of spectral filters and actuators will be beneficial for applications in medical imaging, such as high-throughput screening or point-of-care diagnostics. Future work on the eCLEY will be technologically oriented in order to realize refractive free-form arrays, which are the key to a high optical fill factor and megapixel resolution. A high-volume production at low costs compatible with microelectronics will be feasible, once the investigated mastering technology is an integral part of the microoptical wafer-level technology chain. Hence, the eCLEY offers the smallest total track length at a given resolution and low costs, which are highly attractive to consumer electronics and automotive applications. It was demonstrated to form a fundamentally new approach to miniaturized vision systems with a high resolution that overcomes the basic scaling limits of single-aperture imaging optics.

References

1. M. F. Land and D.-E. Nilsson, *Animal Eyes*, Oxford Animal Biology Series (Oxford University Press, Oxford, 2002).
2. T. Gustavson, *Camera: A History of Photography from Daguerreotype to Digital* (Sterling, New York, 2009).
3. E. Hecht, *Optik* (Addison-Wesley, Bonn, 1994), 3rd edn.
4. W. Vogel, *Glaschemie* (Springer-Verlag, Berlin, 1992).
5. S. Bäumer, *Handbook of Plastic Optics* (Wiley-VCH Verlag, Weinheim, 2005).
6. J. Bareau and P. P. Clark, The optics of miniature digital camera modules, in *International Optical Design, Technical Digest*, CD, G. Groot Gregory, J. M. Howard, and R. J. Koshel, eds (OSA, Washington, DC, 2006), p. WB3.
7. S. Mathur, M. Okincha, and M. Walters, What camera manufacturers want, in 2007 International Image Sensor Workshop (Ogunquit, Maine, 2007).
8. S. Ogata, J. Ishida, and T. Sasano, Optical sensor array in an artificial compound eye, *Optical Engineering* **33**, 3649–3655 (1994).
9. J. S. Sanders and C. E. Halford, Design and analysis of apposition compound eye optical sensors, *Optical Engineering* **34**, 222–235 (1995).

10. K. Hamanaka and H. Koshi, An artificial compound eye using a microlens array and its application to scale-invariant processing, *Optical Review* **3**, 264–268 (1996).
11. K. Hoshino, F. Mura, and I. Shimoyama, Design and performance of a micro-sized biomorphic compound eye with a scanning retina, *IEEE Journal of Microelectromechanical System* **9**, 32–37 (2000).
12. J. Tanida, T. Kumagai, K. Yamada, S. Miyatake, K. Ishida, T. Morimoto, N. Kondou, D. Miyazaki, and Y. Ichioka, Thin observation module by bound optics (TOMBO): Concept and experimental verification, *Applied Optics* **40**, 1806–1813 (2001).
13. M. Shankar, R. Willett, N. Pitsianis, T. Schulz, R. Gibbons, R. T. Kolste, J. Carriere, C. Chen, D. Prather, and D. Brady, Thin infrared imaging systems through multichannel sampling, *Applied Optics* **47**, B1–B10 (2008).
14. G. Druart, N. Guérineau, R. Haidar, S. Thétas, J. Taboury, S. Rommeluère, J. Primot, and M. Fendler, Demonstration of an infrared microcamera inspired by *Xenos peckii* vision, *Applied Optics* **48**, 3368–3374 (2009).
15. J. Duparré, P. Dannberg, P. Schreiber, A. Bräuer, and A. Tünnermann, Artificial apposition compound eye fabricated by micro-optics technology, *Applied Optics* **43**, 4303–4310 (2004).
16. J. Duparré, P. Dannberg, P. Schreiber, A. Bräuer, and A. Tünnermann, Thin compound-eye camera, *Applied Optics* **44**, 2949–2956 (2005).
17. J. Duparré, P. Schreiber, A. Matthes, E. Pshenay-Severin, A. Bräuer, A. Tünnermann, R. Völkel, M. Eisner, and T. Scharf, Microoptical telescope compound eye, *Optical Express* **13**, 889–903 (2005).
18. K.-H. Jeong, J. Kim, and L. P. Lee, Biologically inspired artificial compound eyes, *Science* **312**, 557–561 (2006).
19. J. Duparré and F. Wippermann, Micro-optical artificial compound eyes, *Bioinspiration & Biomimetics* **1**, R1–R16 (2006).
20. D. Radtke, J. Duparré, U. Zeitner, and A. Tünnermann, Laser lithographic fabrication and characterization of a spherical artificial compound eye, *Optical Express* **15**, 3067–3077 (2007).
21. R. Fischer and B. Tadic-Galeb, *Optical System Design* (McGraw-Hill, New York, 2000).
22. M. J. Kidger, *Fundamental Optical Design* (SPIE—The International Society of Optical Engineering, Bellingham, Washington, DC, 2002).
23. R. Kingslake, *Optical System Design* (Academic Press, New York, 1983).
24. R. Kingslake, *A History of the Photographic Lens* (Academic Press, San Diego, CA, 1989).
25. M. Born and E. Wolf, *Principles of Optics* (Cambridge University Press, Cambridge, 1999), 7th edn.
26. H. Naumann and G. Schröder, *Bauelemente der Optik—Taschenbuch der technischen Optik* (Hanser, München, 1992), 6th edn.

27. W. J. Smith, Modern Optical Engineering: The Design of Optical Systems, Optical and Electro-Optical Engineering Series (McGraw-Hill, New York, 1990), 2nd edn.

28. M. Rossi, Wafer-level optics for miniature cameras, in SPIE Photonics Europe—Optics, Photonics and Digital Technologies for Multimedia Applications (SPIE, Brussels, 2010).

29. E. Wolterink and K. Demeyer, WaferOptics mass volume production and reliability, in *Micro-Optics 2010*, vol. **7716**, H. Thienpont, P. V. Daele, J. Mohr, and H. Zappe, eds (SPIE, Brussels, 2010), p. 771614.

30. J. W. Goodman, *Introduction to Fourier Optics* (McGraw-Hill, New York, 1996), 2nd edn.

31. V. N. Mahajan, *Aberration Theory Made Simple* (SPIE Optical Engineering Press, Bellingham, WA, 1991).

32. A. W. Lohmann, Scaling laws for lens systems, *Applied Optics* **28**, 4996–4998 (1989).

33. H. Haferkorn, *Optik* (Wiley-VCH Verlag, Weinheim, 2003), 4th edn.

34. A. Brückner, Microoptical multi aperture imaging systems, PhD thesis, Friedrich-Schiller-Universität Jena (2011).

35. R. Völkel, M. Eisner, and K. J. Weible, Miniaturized imaging systems, *Microelectronic Engineering* **67–68**, 461–472 (2003).

36. J. Duparré, Microoptical artificial compound eyes, PhD thesis, Friedrich-Schiller-Universität Jena (2005).

37. T. Chen, P. Catrysse, A. E. Gamal, and B. Wandell, How small should pixel size be?, in *Sensors and Camera Systems for Scientific, Industrial, and Digital Photography Applications*, vol. **3965**, M. M. Blouke, N. Sampat, G. M. Williams, and T. Yeh, eds (SPIE, San Jose, CA, 2000), pp. 451–459.

38. M. Schöberl, A. Brückner, S. Foessel, and A. Kaup, Photometric limits for digital camera systems, *Journal of Electronic Imaging* **21**, 020501 (2012).

39. F. Wippermann, Chirped refractive microlens arrays, PhD thesis, Technische Universität Ilmenau (2008).

40. F. Wippermann, J. Duparré, P. Schreiber, and P. Dannberg, Design and fabrication of a chirped array of refractive ellipsoidal microlenses for an apposition eye camera objective, in *Optical Design and Engineering II*, vol. **5962**, L. Mazuray and R. Wartmann, eds (SPIE, San Jose, CA, 2005), p. 59622C.

41. J. Duparré, F. Wippermann, P. Dannberg, and A. Reimann, Chirped arrays of refractive ellipsoidal microlenses for aberration correction under oblique incidence, *Optical Express* **13**, 10539–10551 (2005).

42. A. W. Snyder, Acuity of compound eyes: Physical limitations and design, *Journal of Compound Physiology A* **116**, 161–182 (1977).

43. A. W. Snyder, Physics of vision in compound eyes, in *Handbook of Sensory Physiology*, R. Held, H. W. Leibowitz, and H.-L. Teuber, eds (Springer, Berlin, 1977), pp. 225–313.

44. M. C. Hutley, R. Hunt, R. F. Stevens, and P. Savander, The moiré magnifier, *Pure and Applied Optics* **3**, 133–142 (1994).
45. H. Kamal, R. Völkel, and J. Alda, Properties of moiré magnifiers, *Optical Engineering* **37**, 3007–3014 (1998).
46. N. Franceschini, J. M. Pichon, and C. Blanes, From insect vision to robot vision, *Philosophical Transactions of the Royal Society of London B* **337**, 283–294 (1992).
47. J. Kim, K.-H. Jeong, and L. P. Lee, Artificial ommatidia by self-aligned microlenses and waveguides, *Optics Letters* **30**, 5–7 (2005).
48. K. Jeong, J. Kim, and L. P. Lee, Polymeric synthesis of biomimetic artificial compound eyes, in *Proceedings of the 13th International Conference on Solid-State Sensors, Actuators and Microsystems (Transducers 05)*, (2005), pp. 1110–1114.
49. K. Hoshino, F. Mura, and I. Shimoyama, A one–chip scanning retina with an integrated micromechanical scanning actuator, *Journal of Microelectromechanical System* **10**, 492–497 (2001).
50. D. Pätz, S. Leopold, F. Knöbber, S. Sinzinger, M. Hoffmann, and O. Ambacher, Tunable compound eye cameras, in *Micro-Optics 2010*, vol. **7716**, H. Thienpont, P. V. Daele, J. Mohr, and H. Zappe, eds (SPIE, Brussels, Belgium, 2010), p. 77160K.
51. A. Brückner, J. Duparré, P. Dannberg, A. Bräuer, and C. Hoffmann, Ultra-compact vision system for automotive applications, in *Proceedings of 4th EOS Topical Meeting on Advanced Imaging Techniques* (Jena, Germany, 2009), pp. 166–167.
52. A. Brückner, J. Duparré, A. Bräuer, and A. Tünnermann, Artificial compound eye applying hyperacuity, *Optical Express* **14**, 12076–12084 (2006).
53. A. Brückner, J. Duparré, A. Bräuer, and A. Tünnermann, Position detection with hyperacuity using artificial compound eyes, in *Proceedings of SPIE—Sensors, Cameras, and Systems for Scientific/Industrial Applications VIII*, San Jose, CA, vol. **6501**, M. M. Blouke, ed. (IS&T/SPIE, Bellingham, WA, 2007), p. 65010D.
54. J. Thorson, Small-signal analysis of a visual reflex in the locust, *Kybernetik* **3**, 41–66 (1966).
55. K. Nakayama, Biological image motion processing: A review, *Vision Research* **25**, 625–660 (1985).
56. A. Brückner, J. Duparré, P. Dannberg, A. Bräuer, and A. Tünnermann, Artificial neural superposition eye, *Optical Express* **15**, 11922–11933 (2007).
57. E. Buschbeck, B. Ehmer, and R. Hoy, Chunk versus point sampling: Visual imaging in a small insect, *Science* **286**, 1178–1179 (1999).
58. E. K. Buschbeck, The compound lens eye of strepsiptera: Morphological development of larvae and pupae, *Arthropod Structure and Development* **34**, 315–326 (2005).
59. Z. D. Popovich, R. A. Sprague, and G. A. N. Conell, Technique for monolithic fabrication of microlens arrays, *Applied Optics* **27**, 1281–1284 (1988).

60. H. P. Herzig, *Micro-Optics: Elements, Systems and Applications* (Taylor & Francis, London, 1997).

61. S. Sinzinger and J. Jahns, *Microoptics* (Wiley-VCH, Weinheim, 1999).

62. A. Brückner, J. Duparré, R. Leitel, P. Dannberg, A. Bräuer, and A. Tünnermann, Thin wafer-level camera lenses inspired by insect compound eyes, *Optical Express* **18**, 24379–24394 (2010).

63. D. J. Brady and N. Hagen, Multiscale lens design, *Optical Express* **17**, 10659–10674 (2009).

64. A. Oberdörster, A. Brückner, F. C. Wippermann, and A. Bräuer, Correcting distortion and braiding of micro-images from multi-aperture imaging systems, in *Sensors, Cameras, and Systems for Industrial, Scientific, and Consumer Applications XII*, vol. **7875**, R. Widenhorn and V. Nguyen, eds (SPIE, San Francisco, CA, 2010), p. 78750B.

65. A. Oberdörster, A. Brückner, F. Wippermann, A. Bräuer, and H. P. Lensch, Digital focusing and re-focusing with thin multi-aperture cameras, in *SPIE Conference on Electronic Imaging, Digital Photography VIII*, Burlingame, CA, EI116, N. Sampat, F. H. Imai, F. Xiao, S. Battiato, and B. G. Rodricks, eds (SPIE, Bellingham, WA, 2012).

66. P. D. Burns and D. Williams, Practical MTF metrology for digital cameras and scanners, Tutorial notes (Eastman Kodak, San Jose, CA, 2004).

67. OmniVision, OVM7690 640 × 480 CameraCube™ Device, Product brief, 1st edn (2010).

68. G. Druart, N. Guérineau, R. Haïdar, E. Lambert, M. Tauvy, S. Thétas, S. Rommeluère, J. Primot, and J. Deschamps, MULTICAM: A miniature cryogenic camera for infrared detection, in *Micro-Optics 2008*, Strasbourg, vol. **6992**, H. Thienpont, P. Van Daele, J. Mohr, and M. R. Taghizadeh, eds. (SPIE, 2008), p. 69920G.

69. J. Meyer, A. Brückner, R. Leitel, P. Dannberg, A. Bräuer, and A. Tünnermann, Optical cluster eye fabricated on wafer-level, *Optical Express* **19**, 17506–17519 (2011).

70. J. Duparré, D. Radtke, A. Brückner, and A. Bräuer, Latest developments in micro-optical artificial compound eyes: A promising approach for next generation ultracompact machine vision, in *Machine Vision Applications in Industrial Inspection XV*, San Jose, CA, vol. **6503**, F. Meriaudeau and K. S. Niel, eds (SPIE, Bellingham, WA, 2007), p. 65030I.

71. A. Brückner, J. Duparré, F. Wippermann, R. Leitel, P. Dannberg, and A. Bräuer, Ultra-compact close-up microoptical imaging system, in *Current Developments in Lens Design and Optical Engineering XI; and Advances in Thin Film Coatings VI*, San Diego, CA, vol. **7786**, R. B. Johnson, V. N. Mahajan, and S. Thibault, eds (SPIE, Bellingham, WA, 2010), p. 77860A.

72. M. Kawazu and Y. Ogura, Application of gradient-index fiber arrays to copying machines, *Applied Optics* **19**, 1105–1112 (1980).

73. N. F. Borrelli, R. H. Bellman, J. A. Durbin, and W. Lama, Imaging and radiometric properties of microlens arrays, *Applied Optics* **30**, 3633–3642 (1991).

74. E. E. Anderson and W. L. Wang, Novel contact image sensor (CIS) module for compact and lightweight full-page scanner applications, in *Cameras, Scanners, and Image Acquisition Systems*, San Jose, CA, vol. **1901**, H. C. Marz and R. L. Nielsen, eds (SPIE, Bellingham, WA, 1993), pp. 173–181.

75. V. Shaoulov and J. P. Rolland, Design and assessment of microlenslet-array relay optics, *Applied Optics* **42**, 6838–6845 (2003).

76. R. H. Anderson, Close-up imaging of documents and displays with lens arrays, *Applied Optics* **18**, 477–484 (1979).

77. R. Völkel, H. P. Herzig, P. Nussbaum, and R. Dändliker, Microlens array imaging system for photolithography, *Optical Engineering* **35**, 3323–3330 (1996).

78. R. Völkel, H. P. Herzig, P. Nussbaum, P. Blattner, R. Dändliker, E. Cullmann, and W. B. Hugle, New lithographic techniques using microlens arrays, in *MLAs, EOS Top Meeting*, vol. **13**, M. C. Hutley, ed. (EOS, Teddington, 1997), pp. 105–109.

79. R. Berlich, A. Brückner, R. Leitel, and A. Bräuer, Ultra-compact illumination module for multi-aperture imaging systems, in Reliability, Packaging, Testing, and Characterization of MEMS/MOEMS and Nanodevices XI (SPIE MOEMS-MEMS), San Francisco, CA (SPIE, Bellingham, WA, 2012).

80. D. Gabor, Improvements in or relating to optical systems composed of lenticules, UK Patent 541,753 (1940).

81. C. Hembd-Sölner, R. F. Stevens, and M. C. Hutley, Imaging properties of the Gabor Superlens, *Journal of Optics A: Pure and Applied Optics* **1**, 94–102 (1999).

82. M. Hutley, R. F. Stevens, and C. Hembd, Imaging properties of the *Gabor Superlens*, in *Digest of Topical Meeting on Microlens Arrays at NPL, Teddington*, vol. **13**, M. C. Hutley, ed. (EOS, Teddington, 1997), pp. 101–104.

83. K. Stollberg, A. Brückner, J. Duparré, P. Dannberg, A. Bräuer, and A. Tünnermann, The Gabor Superlens as an alternative wafer-level camera approach inspired by superposition compound eyes of nocturnal insects, *Optical Express* **17**, 15747–15759 (2009).

84. S. Farsiu, D. Robinson, M. Elad, and P. Milanfar, Advances and challenges in Super-Resolution, *International Journal of Imaging Systems and Technology* **14**, 47–57 (2004).

85. K. Yamada, J. Tanida, W. Yu, S. Miyatake, K. Ishida, D. Miyazaki, and Y. Ichioka, Fabrication of diffractive microlens array for optoelectronic hybrid information system, in *Proceedings of Diffractive Optics' 99*, (EOS, Jena, 1999), pp. 52–53.

86. J. Tanida, R. Shogenji, Y. Kitamura, K. Yamada, M. Miyamoto, and S. Miyatake, Color imaging with an integrated compound imaging system, *Optical Express* **11**, 2109–2117 (2003).

87. R. Shogenji, Y. Kitamura, K. Yamada, S. Miyatake, and J. Tanida, Multispectral imaging using compact compound optics, *Optical Express* **12**, 1643–1655 (2004).

88. R. Horisaki, S. Irie, Y. Nakao, Y. Ogura, T. Toyoda, Y. Masaki, and J. Tanida, 3D information acquisition using a compound imaging system, in *Optics and Photonics for Information Processing*, San Diego, CA, vol. **6695**, A. A. Awwal, K. M. Iftekharuddin, and B. Javidi, eds (SPIE, Bellingham, WA, 2007), p. 66950F.

89. J. Tanida, Y. Kitamura, K. Yamada, S. Miyatake, M. Miyamoto, T. Morimoto, Y. Masaki, N. Kondou, D. Miyazaki, and Y. Ichioka, Compact image capturing system based on compound imaging and digital reconstruction, in *Micro- and Nano-Optics for Optical Interconnection and Information Processing*, San Diego, CA, vol. **4455**, M. R. Taghizadeh, H. Thienpont, and G. E. Jabbour, eds (SPIE, Bellingham, WA, 2001), pp. 34–41.

90. F. de la Barrière, G. Druart, N. Guérineau, and J. Taboury, Design strategies to simplify and miniaturize imaging systems, *Applied Optics* **50**, 943–951 (2011).

91. R. Lukac, *Single-Sensor Imaging: Methods and Applications for Digital Cameras* (CRC Press, Boca Raton, FL, 2009), Chap. 4, pp. 105–134.

92. Y. Kitamura, R. Shogenji, K. Yamada, S. Miyatake, M. Miyamoto, T. Morimoto, Y. Masaki, et al. Reconstruction of a high-resolution image on a compound-eye image-capturing system, *Applied Optics* **43**, 1719–1727 (2004).

93. K. Nitta, R. Shogenji, S. Miyatake, and J. Tanida, Image reconstruction for thin observation module by bound optics by using the iterative backprojection method, *Applied Optics* **45**, 2893–2900 (2006).

CHAPTER **9**

Future Trends
Panoramic Mini-Cameras

Simon Thibault

Contents

9.1 Introduction

Future trends are always difficult to determine. However, in the field of smart mini-cameras, we can certainly identify panoramic cameras as one such trend. Since the beginning of photography in the 1840s, we have tried to capture the world we see. A panoramic view is one of the best ways to recover the real world on a single image. Consequently, panoramic mini-cameras are of great interest for the future.

Panoramic photography has an old history. At the beginning, photographers created a panorama by making a series of images from a regular photographic camera in which all the images were placed next to each other to create a panoramic image. Figure 9.1

FIGURE 9.1 A panorama composed of six daguerreotypes showing a view of San Francisco, California, in 1853. (From Kingslake, R., *Proceedings of SPIE 0531*, SPIE Press, Bellingham, WA, 1985.)

shows a panoramic view of San Francisco in 1853 (Kingslake 1985). This process can be easily done today using a feature of the digital camera. The real development of panoramic photography was achieved using a curved focal plane. Many cameras used a daguerreotype (a daguerreotype is a direct positive made in the camera on a silvered copper plate) photographic rectangular plate to record wide-angle images up to 150° (horizontal field of view [FOV]). One of the first cameras to use a wide-angle lens to capture a FOV larger than the human eye was in the late 1850s. British photographer Thomas Sutton invented a unique water-filled spherical lens to create panoramic pictures without the need to rotate the camera body. At that time, the Sutton lens was a real improvement as panoramas were very popular, offering a real impression of foreign countries, people, cities, events, and many other exotic things. The fascination with panoramic photography still remains today.

Moreover, the interest in panoramic photography is also motivated by consumers who like to have panoramic videos. However, the increasing trend to use panoramic vision sensors in various applications is driven by the need to have complete information about our surrounding environment. For example, viewing a vehicle's surroundings can directly increase our safety; providing hemispheric vision endoscopic functionality can provide higher patient comfort and a better surgical procedure; and monitoring without a blind zone can provide higher safety in public areas and for home surveillance. Both the increased integration of electronic optical components and the declining prices of electronics in general are possibly the primary enablers of this trend. As the panoramic sensor contributes most to our perception of the world, it is probably one of the most promising ways to fuse many sensors into one, thereby reducing the risk and cost. Consequently, we will focus on the development of panoramic mini-cameras which use a single wide-angle lens to capture the hemispheric FOV rather than multicamera systems.

Section 9.2 describes the development of the first panoramic lens up to the most recent panoramic small lenses. A theoretical background of panoramic lenses is provided in Section 9.3. Section 9.4 presents some of the most recent challenges facing the development of the smart mini panoramic vision sensor. Finally, in Section 9.5, some recent researches which can help the development of a miniature panoramic imager are discussed.

9.2 Road to Panoramic Smart Camera

In the late 1800s, fairly large panoramic cameras were necessary to accommodate the large size, curved photographic plates (Kingslake 1985). Distortion was also a concern. At that time, pictures were recorded on photographic plates. Using both a limited FOV and a curved photographic plate has reduced the distortion impact. However, with the increase in the development of lens design techniques and materials, the idea of developing a lens with a hemispheric FOV became a reality. To project a hemispheric FOV onto a flat detector plane (photographic film or focal plane array), we need a large barrel distortion. However, this distortion should not be considered an aberration but rather the result of the projection of a hemispheric field on a circle, which is not possible without distortion. This type of lens is called a fisheye lens.

The classical example of a fish-eye lens "type" of image formation is an actual fish eye under water (Miyamoto 1964; Kumler and Bauer 2000). In 1911, Robert W. Wood described a water-filled pinhole camera that was capable of simulating a fish's view of the world. The water was not practical; however, Bond added a hemispheric lens with a pupil at the center of the curvature to obtain the same effect without water. In 1924, Hill developed his Sky lens by adding a diverging meniscus lens before the hemispheric lens to improve the field curvature (thereby reducing the Peztval sum). This lens was the first prototype of the modern fish-eye lenses which were patented by Schultz (1932) and Merté (1935). Some 40 years later, the well-known afocal wide-angle door viewer was patented (Hiyashi 1978).

The fish-eye lens suffers from severe drawbacks particularly when the fish eye is facing up or down. In these positions the subject of interest might appear at the edge (large angle) of the FOV where the barrel distortion is very large. The fish-eye lens has a linear relation between the FOV and the image height. This linear relation produces a constant magnification or a constant angular resolution (pixel per degrees or pixel per milliradians [mrad]). For example, for surveillance (indoor), a camera is usually mounted on a ceiling, with its lens facing down, as shown in Figure 9.2. The most significant objects are in the zone at the periphery of the lens. This part of the picture is the most important because it enables facial recognition or detection. To develop a smart imaging system, we would like to have a higher magnification in the zone at the periphery (zone of interest [ZoI]). Consequently, the fish-eye image distortion is a limitation in many applications.

The fish-eye lens also produces a circular image footprint, as shown in Figure 9.3, which is not an optimal solution. Indeed, on a rectangular detector, the fish-eye footprint is not optimal because it will cover less than 60% of the sensor pixel.

To solve these issues, a new design technique was developed in 2005, to manage the distortion and the pixel coverage for wide-angle hemispheric lenses (Thibault 2005). Distortion management offers a new degree of freedom for optical designers. With some imagination, the qualified specialist can design optical systems that enhance the performance of digital panoramic imaging systems. By controlling distortion, the relation between the FOV and the pixels may be optimized for any given application. To maximize the optical performance, the lens may be designed to increase the magnification in a ZoI. This is done by controlling the slope of the distortion as a function of the FOV. Figure 9.4 shows the pixel coverage obtained by proper distortion control in Figure 9.2 with a ZoI in Zone B. The dashed line represents the

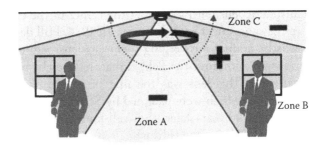

FIGURE 9.2   A camera is mounted on a ceiling in an indoor surveillance application. The plus sign shows the zone where higher magnification is required.

FIGURE 9.3   Fish-eye image footprints on a rectangular sensor. The black squares show the constant magnification.

pixel coverage for an ideal fish-eye lens, which is constant with the field angle. Moreover, we can also add cylindrical power to the lens in order to obtain a more convenient image footprint which covers up to 80% of the sensor pixels. Figure 9.5 shows the image footprint given by a lens with increased magnification in the ZoI at the periphery as well as an elliptical image coverage with a 4:3 ratio. Thus, managing distortion and pixel coverage is an innovative approach that can be used in smart panoramic imagers. This new type of lens is called the panomorph lens.

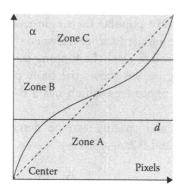

FIGURE 9.4   Pixel coverage obtained by proper distortion control. The dashed line represents the pixel coverage for an ideal fish-eye lens, which is constant with the field angle.

FIGURE 9.5 Panomorph image footprint on a rectangular sensor. The black squares show the variation of the magnification.

9.3 ■ Panoramic Lens: Theoretical Background

This section describes the theoretical background regarding the image formation from a panoramic lens. The panoramic refractive lens is based on the design of a reversed telephoto lens. The reversed telephoto lens has a back focal length (BFL) which is longer than the effective focal length (EFL). This is achieved using a negative lens at the front focal point of an objective (positive) lens or camera lens. The back focus distance is lengthened while the EFL is unchanged. These lens combinations were very popular for single-lens reflex (SLR) 35 mm cameras where a long distance between the last element and the film was required to clear the flipping mirror when an exposure was made. Figure 9.6 shows the basic power arrangement for a reversed telephoto lens which has a BFL that is longer than the EFL.

The power and spacing can be determined by the simple Equations 9.1 and 9.2:

$$F_a = \frac{D \cdot \text{EFL}}{\text{EFL} - \text{BFL}} \qquad (9.1)$$

$$F_b = \frac{-D \cdot \text{BFL}}{\text{EFL} - \text{BFL} - D} \qquad (9.2)$$

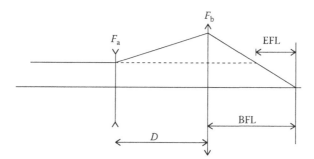

FIGURE 9.6 Basic power arrangement for a reversed telephoto lens which has a back focal length that is longer than the effective focal length.

The BFL can be very long but the Petzval field curvature has a tendency to become overcorrected (backward curved). Consequently, the optimum arrangement to compensate for the Petzval curvature is when the focal lengths of the two components are nearly equal (opposite sign).

When the aforementioned reverse telephoto lens is designed in an extreme form (short focal length), a hemispheric FOV or larger can be obtained. The front element becomes a strongly bended negative meniscus with the concave surface toward the image. Additionally, the size of the front element increases compared to the entrance pupil dimension. This negative meniscus serves to bend the ray coming from a large angle through the stop. A large amount of distortion is then introduced but it is absolutely required to image the hemispheric FOV on a finite-size detector. For such high-distortion lenses, the standard focal length to image height relation is no longer relevant ($h' = f\tan\theta$). Consequently, the wide-angle lens will have an image–height relation according to $h' = f\theta$ or other types of projection equations (Miyamoto 1964). The linear relation is the one used by a fish-eye lens which will produce a constant magnification within the FOV (the derivative of the linear function by the angle will be constant).

As discussed in the last section, a smart panoramic lens will include a variation within the FOV which will produce a magnification variation in the ZoI. Then, the FOV–image relation will be given by a polynomial function (Equation 9.3).

$$H = a + b\theta + c\theta^2 + d\theta^3 + \cdots \tag{9.3}$$

By definition, *a* is equal to zero (no dc value) and *b* is the EFL. However, the nonzero value of *c*, *d*, and higher-order terms will introduce a variation in the projection of the FOV in the image plane. In this case, we will use the local focal length to express the variation of the magnification. From a geometrical optics point of view, as the object and the image distance are constant, the way to modify the magnification is to change the focal length. Equation 9.4 gives the local focal length expression:

$$f(\theta) = \frac{\partial H}{\partial \theta} = b + c\theta + d\theta^2 + \cdots \qquad (9.4)$$

The localized focal length $f(\theta)$ is expressed in millimeters when the angle is in radians, but it is sometimes more convenient to use millimeters per degree or pixels per degree, which correspond to the inverse of the instant FOV (per IFOV), sometimes called the plate scale. In panoramic applications, we refer sometimes to the resolution as the number of pixels per degree. We understand here that the resolution is not the optical resolution which expresses the image quality, such as the modulation transfer function (MTF).

Figure 9.7 shows the impact of the localized focal length on an image. The visual impact is the magnification variation which follows the variation of the focal length of the imager.

FIGURE 9.7 The top picture shows an imager with a longer focal length in the central part. The bottom figure shows an imager with a longer focal length on the edge.

9.4 Challenges: Lenses, Sensors, and Processing

As a new technology, the design of a smart, panorama-based visual sensor requires particular attention. The following section discusses a number of challenges that illustrate the key points to be considered in designing such a vision sensor. These challenges are briefly discussed in terms of lens design and material, sensor, and processing.

9.4.1 Lenses

Currently, wide-angle smart lenses exist on the market. Over the last few years, it has been demonstrated that wide-angle lenses, while more challenging, can be built to replace a regular lens in various applications. From an optical design point of view, we can consider some aspects that are unique to miniature applications (as discussed in Chapter 8). The scale camera system creates unique challenges for lens designers. When designing such a wide-angle lens, it is not always helpful to use a traditional larger-scale lens as a starting point. Scaling down the lens will result in elements that cannot be built. For example, the element thickness can become too thin to be fabricated. The challenge is to maintain a high precision, compact lens design using glass and plastic elements.

The development of sensors has been moving toward smaller pixels and higher-density formats. The higher-resolution formats have made the lens designer's job extremely challenging because the reduction of the pixel size has required an increase in the lens performance. The lens performance requirements are closer to the diffraction limits and the lens tends to be more sensitive to misalignment. Another problem with smaller pixels is their limited light collection ability. The sensor surface is not uniformly sensitive. Circuitry integrated with the sensor reduces the active area significantly. To improve sensitivity, an array of microlenses located above the active area of the sensor is applied but this will also impose a chief ray angle (CRA) specification on the lens design. The CRA is the angle of incidence of the chief ray at the image plane for any field point. The microlens acts as a condenser, relaying the sensor image to the exit pupil of the microlens. This increases the apparent pixel size.

The total track length (TTL) of the lens (from first optical surface to sensor) is also an important factor for miniature optics. The TTL of mobile-phone cameras on the market can be as small as 2 mm. Until now, the smallest wide-angle lens TTL was about 5.5 mm (Immervision Panomorph lens, www.immervision.com).

In general, a wide-angle lens is about two times longer than a 30° lens. However, new optical techniques, such as new plastic materials, nanostructured patterns, and wafer-scale optics, look promising in the design of short TTL panoramic lenses.

As part of the design process, the material selection is critical. Plastic injection-molded optics can be a good choice. Keeping in mind its manufacturing limits, plastic has the big advantage that the flanges can be molded to eliminate the need for spacers and allow for mechanically driven centering of one element to another. One disadvantage is that there are very few plastic materials (low and limited index of refraction), so the choices are limited. The temperature requirement can also rule out some types of regular plastic, such as those used in mobile-phone cameras. Another option that is available is molded glasses, which allow the advantages of both a high index and aspheric correction. The manufacturing process is more complicated and less flexible than plastic; nevertheless, molded glass can be the material of choice when the goals are stability, color correction, and extreme temperature conditions, as in the automotive industry.

The lens performance for digital vision systems is commonly expressed in terms of MTF at spatial frequencies between Nyquist/2 (Ny/2) and Nyquist/4 (Ny/4). The Nyquist frequency is $(2 \times \text{pixel size})^{-1}$. So, for a 5.6 μm pixel, Ny/2 is 45 lp/mm; for a 2.2 μm, Ny/2 is 113.6 lp/mm; and for a 1.75 μm, Ny/2 is 142.9 lp/mm. The size of the sensor is not as critical as the pixel pitch for the design of a panoramic lens. The image quality must be as uniform as possible because the center and the edge are as important in a panoramic imaging system.

9.4.2 Sensors

Balancing sensor parameters to handle better imaging performances is also key for smart panoramic imaging systems. To design the best imaging system, the sensor performance, such as the sensitivity, the speed, the noise, and the compression format, must be optimized. Consequently, the lens design itself is not enough; the sensor must be tuned to a higher level to have a smart imaging system.

One aspect of the panoramic image is the black zone all around the imaging areas. As the panoramic image is captured as a circle or an ellipse, the corners of the sensor are not illuminated. This nonilluminated area will have an impact on the white balance and the gain level. Having 30% or more of the

FIGURE 9.8 Real-time distortion-free display (*left*: original image produced by the panomorph lens).

sensor not illuminated will affect the imaging performance of the sensor. A smart panoramic lens such as a panomorph lens will reduce the nonilluminated areas (see Figure 9.8). However, some pixels will still not be illuminated. Consequently, the black areas around the image footprint will impact on the overall performance. The impact of this effect has not yet been documented but research should continue to develop an optimized smart panoramic imager.

9.4.3 Processing

A panoramic lens has inherent distortion; however, this distortion, discussed earlier, should not be considered as an aberration but as a result of the projection of a hemisphere on a plane. Considering the angle of incidence θ (in radian) of a light coming from an object at a long distance, the coordinates in the image plane will be (u,v). The lens will image the object as a function of the angle θ. This function can be linear but not necessarily $(u = v = \text{constant} = f \times \theta$ for an ideal fish eye). In the case of a smart panoramic lens such as a panomorph lens, the relation between u and v is proportional to the anamorphic ratio and to the polynome within θ (Equation 9.3). The variation of the local focal length across the FOV is the main advantage. The derivative of u or v with respect to θ is the lens resolution (according to the definition given at the beginning of the chapter). For a fish eye, the resolution is constant (ideal fish eye), but for a panomorph lens, the resolution is also a polynomial function with θ.

To be effective, the panoramic image must be displayed under a distortion-free image in real time. Consequently, the viewing process must unwrap the image in real time in order to provide views that reproduce real-world proportions and geometric information. The algorithms should be customized and adapted for each specific application, which is then related to human vision (display) or artificial vision (analytic function).

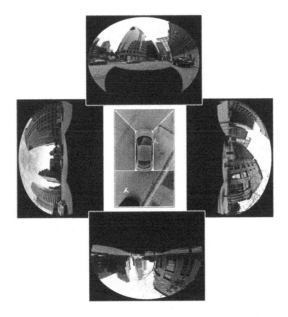

FIGURE 9.9   View of the area surrounding a vehicle using four panomorph lenses.

For example, for an automotive application, four smart panomorph lenses could be mounted all around a vehicle to provide a complete view with added benefits (see Figure 9.9).

The software must also be as light as possible to be embedded into several platforms. A smart imager is also demanding on software performances. This is not because the memory is limited, but rather it is a question of speed. Long processing will impact the frame rate.

9.5 Research Trends

In this section, we will briefly discuss future research trends. The goal is not to cover all aspects but rather to give the reader some insight into active research areas.

The quest for a miniature optical system is highly driven by the consumer electronic market, which requires a smaller and smaller optical form factor with a higher resolution. Medical imaging is also another field where miniaturization is important. This a particular challenge for panoramic lenses. Recently, Milojkovic and Mait (2012) explored how changes in optical design affect the imaging performance as a function of the system size. He explored not only the lens composition but also the curved detector and multiple detectors. He suggested that monocentric lenses combined with a curved detector are an optimal solution for a higher space-bandwidth. This work also shows that the conclusion of such a study is highly related to the original criteria of comparison. Moreover, Milojkovic barely studied the performance of graded index Luneberg-type lenses or the impact on lens' cost and size. Consequently, improvements on various fronts are still to be made in the field of miniature panoramic imaging systems. In particular, new optical components that combine nanowire material to design achromatic lenses (Costa and Silveirinha 2012) are interesting approaches for achromatic miniaturization. As an achromatic lens is difficult to miniaturize, the approach looks promising for the coming years. Transformation optics and metamaterials also offer great potential and versatility for controlling aberrations (Chen et al. 2010; Fuerschbach et al. 2011). However, commercial optical design software does not provide the tools to use such new technologies to design optical imagers. Further development will be required to use these new technologies within a real lens design. This area of research can be a real breakthrough for future applications and the development of miniature panoramic lenses.

During the last few years, the segmented field lens technique (microlens and macrolens arrays) has revealed its power in many applications to miniaturize an imager. Integral field spectrometers use an array of lenslets to integrate the light within subregions of the image and the energy is then dispersed by a prism. The Shack–Hartman wave-front sensor uses a lenslet array to analyze the wave-front component. The lenslet is used to concentrate the light and to measure the image point displacement induced by the wave-front gradient. In both cases, the lenslet is not used to correct the aberration of the lens system. By controlling the lenslet shape, we can potentially correct the local wave front on the lenslet. In addition, the first paper on this subject was recently published by Brady and Hagen (2009). Brady's paper presents the basic ideas of the multiscale lens whose goal is to correct the local wave front (small errors) rather than the global wave-front error at the image plane. The resulting image quality would then be higher than the conventional design. Much work is still required to make this technique viable.

Finally, we have to speak about the combination of hardware and software to design new smart imaging systems. Wave-front coding is a technique developed to enhance the depth of focus of imaging systems, but it can also be used to reduce focus-related aberrations (Dowski and Johnson 1999; Mezouari and Harvey 2003; Mezouari et al. 2006) and ease tolerancing (Yan and Zhang 2008; Lee et al. 2010). The basic concept of this method is to extend the depth of focus by inserting a phase mask near the system aperture stop. This phase mask degrades the MTF but in a way that it can easily be reconstructed by post-processing. The efficiency of this technique has been shown in theory and real-world applications (Feng et al. 2010; Muyo and Harvey 2004). However, the application of this technique to wide-angle systems poses some challenges. In order to perform a proper image reconstruction, the point spread function (PSF) must be sufficiently constant over the whole captured image. The wide-angle lens has varying off-axis aberrations. Consequently, the PSF is not uniform, which makes wave-front coding difficult to apply. Recently, research has demonstrated, at least on paper, that the phase mask parameters can be used to mitigate the off-axis differential aberrations (Lariviere-Bastien et al. 2011). Instead of correcting for defocus, the wave-front coding will be used to correct for axis aberrations. The goal is to develop a much simpler panoramic lens that can be corrected by a proper phase mask. A simpler panoramic lens means a limited number of elements. Less optical elements will lead to a more compact lens design, a shorter TTL.

9.6 Concluding Remarks

The panoramic camera is only in its early stages and it will become increasingly relevant in our lives. This is mainly because the panoramic lens will produce an image that is closer to what our eyes see. Consequently, panoramic cameras will be very popular in the near future.

Using actual technologies, lens designers can design a wide-angle lens to some extent. With the development of new technologies, such as the curved detector and transforming optics, the dream of a miniature panoramic lens on a mobile phone will be possible within the next 5 years.

References

Brady, D.J. and Hagen, N. (2009), Multiscale lens design, *Opt. Express* *13*, 10659–10674.

Chen, H., Chan, C.T., and Sheng, P. (2010), Transformation optics and metamaterials, *Nat. Mater.* *9*, 387–396.

Costa, J.T. and Silveirinha, M.G. (2012), Achromatic lens based on a nanowire material with anomalous dispersion, *Opt. Express 20*, 13915–13922.

Dowski, E.R. and Johnson, G.E. (1999), Wavefront coding: A modern method of achieving high performance and/or low cost imaging systems, *Proceedings of the Current Developments in Optical Design and Optical Engineering VIII*, 137–145.

Feng, L., Meng, J., and Dun, X. (2010), The application of wavefront coding technology to a large segmented synthetic aperture telescope, *Proceedings of SPIE 7654*, SPIE Press, Bellingham, WA, p. 76540E.

Fuerschbach, K., Rolland, J.P., and Thompson, K.P. (2011), A new family of optical systems employing φ-polynomial surfaces, *Opt. Express 19*, 21919–21928.

Hiyashi, T. (1978), Wide-angle optical system for door viewer, U.S. Patent 4,082,434.

Kingslake, R. (1985), Development of the photographic objective, *Proceedings of SPIE 0531*, SPIE Press, Bellingham, WA, pp. 60–67.

Kumler, J. and Bauer, M. (2000), Fisheye lens design and their relative performance, *Proceedings of SPIE 4093*, SPIE Press, Bellingham, WA, pp. 360–369.

Larivière-Bastien, M., Zhang, H., and Thibault, S. (2011), Distributed wavefront coding for wide angle imaging system, *Proceedings of SPIE 8128*, SPIE Press, Bellingham, WA, pp. 81280G–81280G-8.

Lee, S.-H., Park, N.-C., and Park, Y.-C. (2010), Improving the tolerance characteristics of small F/number compact camera module using wavefront coding, *Microsyst. Technol. 16*(1–2), 195–203.

Merté, W. (1935), D. R. Patent No. 672393.

Mezouari, S. and Harvey, A.R. (2003), Phase functions for the reduction of defocus and spherical aberration, *Opt. Lett. 28*, 771–773.

Mezouari, S., Muyo, G., and Harvey, A.R. (2006), Circularly symmetric phase filters for control of primary third-order aberrations: Coma and astigmatism, *J. Opt. Soc. Am. A 23*, 1058–1062.

Milojkovic, P. and Mait, J.N. (2012), Space-bandwidth scaling for wide field-of-view imaging, *Appl. Opt. 51*, A36–A47.

Miyamoto, K. (1964), Fish eye lens, letters to the editor. *JOSA 54*, 1060–1061.

Muyo, G. and Harvey, A.R. (2004), Wavefront coding for athermalization of infrared imaging systems, *Proceedings of SPIE 5612*, SPIE Press, Bellingham, WA, pp. 227–235.

Schulz, H. (1932), D. R. Patent No. 620538.

Thibault, S. (2005), Distortion control offers optical system design a new degree of freedom, *Photon. Spectra 5*, 80–82.

Yan, F. and Zhang, X. (2008), The effect on tolerance distributing of an off-axis three mirror anastigmatic optical system with wavefront coding technology, *Proceedings of SPIE 7068*, SPIE Press, Bellingham, WA, pp. 706807–706814.

Index

Printed in the United States
by Baker & Taylor Publisher Services